Distillation Topical Conference 2010

Topical Conference at the 2010 AIChE Spring Meeting and 6th Global Congress on Process Safety

San Antonio, Texas, USA
21 – 25 March 2010

ISBN: 978-1-61738-443-1

Printed from e-media with permission by:

Curran Associates, Inc.
57 Morehouse Lane
Red Hook, NY 12571

CURRAN ASSOCIATES INC. proceedings.com

Some format issues inherent in the e-media version may also appear in this print version.

Printed by Curran Associates, Inc. (2010)

For permission requests, please contact AIChE
at the address below.

AIChE
3 Park Avenue
New York, NY 10016-5991

Phone: (203) 702-7660
Fax: (203) 775-5177

www.aiche.org

Additional copies of this publication are available from:

Curran Associates, Inc.
57 Morehouse Lane
Red Hook, NY 12571 USA
Phone: 845-758-0400
Fax: 845-758-2634
Email: curran@proceedings.com
Web: www.proceedings.com

TABLE OF CONTENTS

(31a) Column Profiles for Membrane Column Sections

Mark Peters, Wits University, Johannesburg, South Africa

Diane Hildebrandt, Comps, Wits University, Johannesburg, South Africa

David Glasser, Comps, Wits University, Johannesburg, South Africa

Shehzaad Kauchali, Wits, Wits University, Johannesburg, South Africa

A novel graphical method of analysing continuous membrane separation systems has been developed. The method is applicable to all counter- and co-current membrane modules – single-stage, cascades, and membrane columns. All configurations can be broken down into "column sections", no matter how complex the arrangement. Each column section, which resembles a typical single-stage separator, is modelled using the difference point equation. The difference point equation tracks the change in the retentate composition down the length of the column section. For demonstration purposes, a simple constant relative permeability flux model is employed to determine the permeate composition as well as the continually changing retentate flow. All possible operating conditions of a column section are explored. Each condition is a result of the relationships between the flows and compositions at the top of a column section. Using the difference point equation, column profiles for each condition can be plotted. The behaviour of the profiles is discussed both mathematically and graphically. This innovative way of investigating membrane processes provides a unique way of synthesizing and designing them.

See more of this Session: Advances in Distillation & Absorption I
See more of this Group/Topical: Topical 8: Distillation Topical

(31b) Methodology and Development of a Process for Extractive Distillation Using a Novel Ionic Liquid for the Recovery of Olefin From Paraffin Mixtures

Gary Belluomini, Engineering Sciences, The Dow Chemical Company, Freeport, TX

John Pendergast, Engineering Sciences, The Dow Chemical Company, Freeport, TX

Jacinto Lopez-Toledo, Engineering Sciences, The Dow Chemical Company, Freeport, TX

Yujun Liu, Engineering Sciences, The Dow Chemical Company, Freeport, TX

Mahesh Sawant, Engineering Sciences, The Dow Chemical Company, Freeport, TX

Chris Domke, Engineering Sciences, The Dow Chemical Company, Freeport, TX

A methodology and a proposal for the design of an extractive distillation process based upon using a novel ionic liquid as the heavy solvent to separate olefins from paraffins is presented. The ionic liquid is silver(I)/N,N-dimethylbenzoamide bis(trifluoromethylsulfonyl) imide [$Ag(DMBA)_2$+Tf_2N-], and was synthesized by the Oak Ridge National Laboratory. This ionic liquid takes advantage of the special ability of silver to form pi-bonds with the double bonds in the olefin. The process is designed using the ASPEN PLUS™ process simulation tool, and the steps that are required for the process development are described.

The application that will be discussed is the recovery of higher olefins from paraffins. In this case, the discussion will center on the separation of 1-octene from the binary mixture of n-octane and 1-octene. The model is based on experimental infinite dilution data that were generated and recently published by The Dow Chemical Company in the Industrial & Engineering Chemistry Research journal. However, the steps and data requirements in the development of the process are broadly applicable to the development of a broad variety of processes using ionic liquids as the heavy solvent in an extractive distillation process.

The development of the information and the evaluation of the potential benefits of using ionic liquids is a necessary step for these relatively new class of components to gain wide commercial acceptance. Any broad application of these materials must involve an understanding of the economic benefits and economic hurdles yet to be overcome. In an effort to understand the economic impact of an extractive distillation using an ionic liquid, the proposed extractive distillation scheme using ionic liquid is compared with a conventional process using traditional heavy solvent like N-methyl pyrrolidone.

See more of this Session: Advances in Distillation & Absorption I
See more of this Group/Topical: Topical 8: Distillation Topical

(31c) De-Bottleneck a PP Splitter with 4 Pass VGPlus Trays

Giuseppe Mosca, Technology, Sulzer Chemtech, Milano, Italy

Elena Tacchini, Technology, Sulzer Chemtech, Milano, Italy

Propylene is especially important for the production of the most varied industrial and consumer goods such as fittings and packaging material, as well as injection moldings of plastics and many others. It has been estimated that the growth rate of world Propylene consumption will substantially exceed the one of Ethylene consumption. Among all the emergent countries, Russia is one of the those in which the Propylene demand is expected to grow the most. In the last few years several Steam Crackers (Ethylene Plants) have been revamped in Russia to meet this request. Among all the distillation columns of an Ethylene Plant, the Propylene-Propane Splitter is surely the most critical tower. This is even more true in many Russian Steam Crackers where the PP-Splitters were equipped with "non conventional fractionation trays", providing with poor performances, therefore the recovery of Propylene was much lower than commonly achieved in modern Plants. In this paper a case story will be presented: the revamp of the PP-Splitter with the most updated mass transfer technology allowed for a significant boosting of the Propylene recovery. It will include a detailed discussion of the revamp approach: the process simulation, the selection of the most appropriate fractionation trays, the job implementation at site, and the achieved results.

See more of this Session: Advances in Distillation & Absorption I

See more of this Group/Topical: Topical 8: Distillation Topical

DETECTING INCIPIENT FLOODING IN A DISTILLATION COLUMN

Roger K. Pihlaja – Emerson Process Management, Rosemount, Inc., Chanhassen, MN

Biography

Roger Pihlaja works for the Rosemount Measurement Division of Emerson Process Management, in the area of advanced pressure diagnostics applications. His career at Emerson has involved developing diagnostics applications in a wide range of process industries. Prior to Emerson, Roger worked for about 26 years in process development at the Dow Chemical Co. Roger received BS and MS degrees in chemical engineering from Michigan Technological University. Roger also earned a BS degree in mechanical engineering from Saginaw Valley State University. He has five issued US patents and numerous other pending applications.

Abstract

Distillation is a common unit operation, which is used to separate or purify components in a feed stream. Flooding is a common abnormal process condition wherein the distillation column stops generating a separation. This presentation will document experiments with the on-board Advanced Diagnostics option in the Rosemount 3051S differential pressure (DP) transmitter to detect Incipient Flooding on-line in real time. A 3051S DP transmitter was installed across the packed section of the 18" diameter distillation column at the UT, Austin, Separations Research Program facility. The on-board signal processing capability of the Advanced Diagnostics option was used to detect the change in gas flow regime at the onset of flooding and generate a robust Incipient Flooding alarm. There are thousands of distillation columns in service. Because this technique detects a physical phenomena intrinsically associated with flooding, it is believed to be widely applicable. Users of Rosemount DP transmitters will be able to confidently push their distillation columns closer to their operational limits and quickly eliminate flooding as a root cause of product quality issues.

Introduction – How Distillation Columns Work

Distillation is a process used across many industries including refining and chemical. In a distillation system a feed stream containing a mixture of components is separated or purified into two or more product streams enriched in certain feed components and depleted in others. Distillation is probably the most common separation or purification process in use today.[1] The process of distillation is very energy intensive and can contribute to more than 50% of plant operating costs.[1] Thus, there is great interest in operating distillation columns as efficiently and reliably as possible. Figure 1 illustrates the basic components of a distillation system.[2]

A feed stream, usually liquid, consisting of a mixture of components to be separated, is fed into a vessel. This vessel is usually a tall cylinder in shape, standing vertically, and is

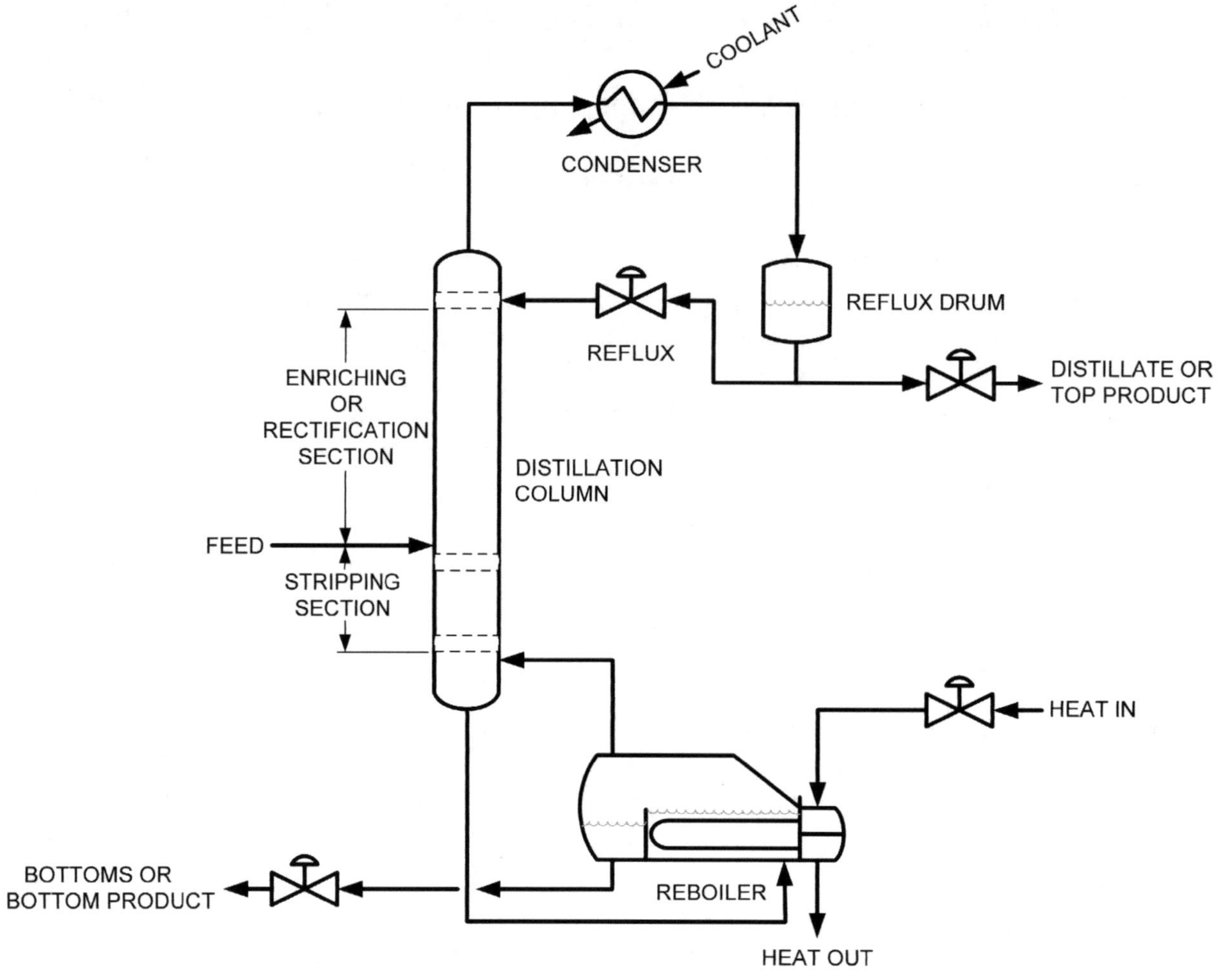

Figure 1 - Basic Components of Distillation

commonly referred to as a "distillation column."

Inside the distillation column are structures designed to cause intimate radial mixing (i.e. mixing at any given vertical level); but, not axial mixing (i.e. mixing along the length of the vessel). These structures also provide contact area between a stream of vapor flowing up and a stream of liquid flowing down. These structures can take the form of discrete "trays" or "stages" or they can be in the form of packing. Packing can be randomly dumped shapes such as "Rashig Rings", "Beryl Saddles", etc. Or, the packing can be a so-called "structured packing" such as "Koch Flexipack^{tm}."

The purpose of all these different types of column internal structures is the same: to facilitate vapor/liquid contacting and mass transfer. The part of the distillation column above the feed point is commonly referred to as the "Enriching Section" or the "Rectification Section." The part of the distillation column below the feed point is known as the "Stripping Section."

Liquid trickles down the column, exits the bottom of the Stripping Section, and flows into the reboiler. The reboiler is a special type of heat exchanger that uses steam or some other heat transfer fluid to heat the liquid in the reboiler to its boiling point. The vapor generated by this boiling liquid exits the reboiler and is fed back into the Stripping Section of the column. Excess liquid in the reboiler overflows a weir and exits the process as the "Bottoms Product", sometimes referred to as the "Bottoms".

The vapor from the reboiler flows up the column, countercurrent to the liquid flowing down the column. The components in the feed stream are separated according to their relative boiling points. Components with a lower boiling point tend to become enriched in the vapor traveling up the column. Components with a higher boiling point tend to become enriched in the liquid traveling down the column. Eventually, the vapor enriched in low boiling components exits the Rectification Section on the top of the distillation column. This vapor is condensed back to a liquid by cooling in a heat exchanger called the condenser. The condensed liquid is collected in the reflux drum. A portion of the condensed liquid is fed back into the Rectification Section to become the liquid flowing down the column as described previously. The rest of the liquid exits the process as the "Top Product" or the "Distillate".

Flooding in Distillation Columns

A common problem that can occur in distillation columns is flooding. Figure 2 illustrates a cross section through a distillation column, filled with a structured packing, showing the liquid flow and vapor flow when the column is operating normally. Liquid is flowing downward over the structured packing countercurrent to the upward flowing vapor. The vapor must follow a tortuous path, but the void space in the packing is predominantly filled with vapor. The vapor is said to be the "continuous phase". The upward flow of the vapor exerts an "aerodynamic drag" on the falling liquid. This drag force acts in opposition to the force of gravity and slows the flow of the falling liquid. When the relative flow rates of the vapor and liquid are such that the drag force is greater than or equal to the gravity force, then the liquid stops flowing down the column. This condition is called flooding. Flooding can begin at any vertical location in the column.

Figure 3 illustrates the same cross section through the column as Figure 2 except the column is flooding. Note how the flow regime has changed. The liquid is now the continuous phase with bubbles of vapor rising through it. The rising bubbles tend to drag a lot of liquid upward, thus causing undesirable axial mixing in the column. The net effect is the distillation column stops generating a separation and the top product and bottom product become similar in composition to the feed stream. The bubbles have a distribution of sizes and tend to nucleate, grow, agglomerate, and break apart randomly. As a result, bubble flow tends to be a random, chaotic process. If the pressure drop were measured across the packing, it is expected this random bubble flow process would contribute a broad spectrum or "white noise" component to the pressure drop signal. This white noise component is not expected to be present during normal operation. Thus, it should be a good diagnostic signal to detect flooding.

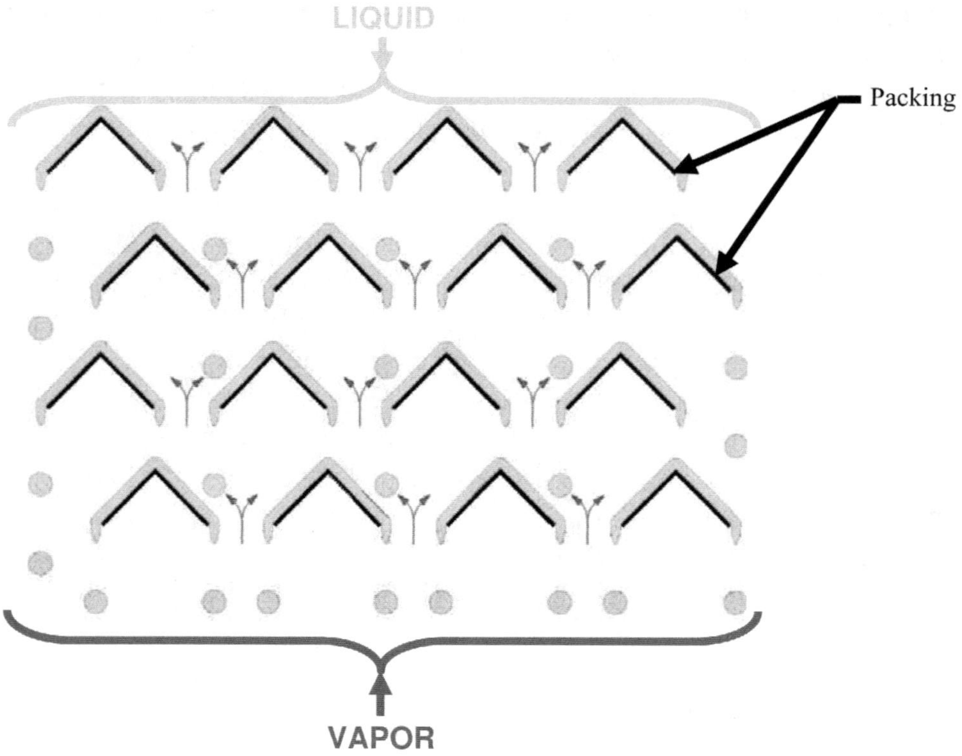

Figure 2 - Normal Operation: Vapor Flow Is Continuous Phase With Falling Liquid Droplets

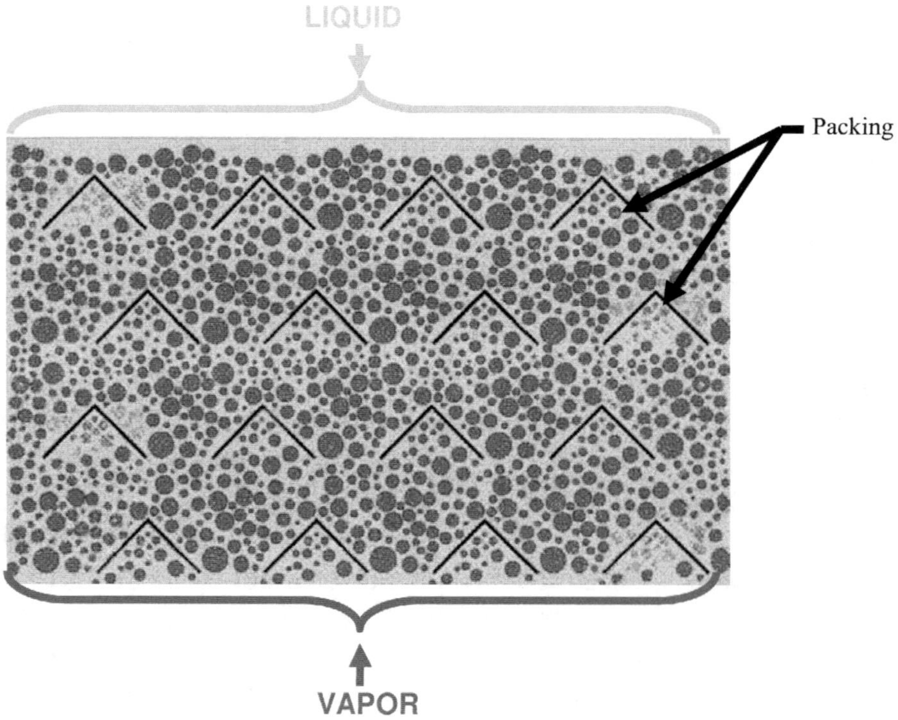

Figure 3 - Flooded Operation: Vapor Flow Becomes Rising Bubbles in Continuous Liquid Phase

Figure 4 shows the conditions visible through a sight glass in a distillation column filled with a structured packing in normal and flooded operation. In Figure 4A, with the column in normal operation, note the open void space within the packing with liquid weeping out of the packing at several points.

A - Column in Normal Operation **B – Column in Flooded Operation**

Figure 4 – View Port of Operating Distillation Column

Now, contrast Figure 4A with the view through the same sight glass when the column is flooding in Figure 4B. The highly chaotic bubble flow is quite evident. A motion picture of the fluid action in Figure 4B would resemble the view through the viewing port on a front loading clothes washing machine.

In actual practice, distillation columns usually have numerous control loops, running in Auto mode, controlling process parameters like feed rate, column pressure, reboiler heat duty, reflux ratio, top and/or bottom product compositions, various temperatures, etc. The action and interaction of these control loops can make detection and diagnosis of flooding difficult.[3] In order to make it easier to study flooding in distillation columns, it is common practice to simplify the operation of the system. Figure 5 illustrates how this is accomplished by eliminating the feed stream into the column as well as the distillate product and bottom product streams out of the column. Instead, all the condensed top product is sent back down the column as liquid reflux. All the bottom product is boiled back up the column as vapor by the reboiler. Distillation columns operating in this mode are said to be on “total reflux” or “100% reflux mode”. In 100% reflux mode, the only process parameter that can be changed is the reboiler heat duty. Flooding is triggered by increasing the reboiler heat duty. Increasing the reboiler heat duty causes more vapor to be boiled up, which increases the vapor flow up the column. Eventually, the increasing vapor flow rate exceeds a critical threshold and triggers flooding.

Although Figures 2 - 4 tend to imply flooding is simply a discrete On/Off phenomenon, this is not necessarily true. In a typical distillation column, as the relative vapor velocity is increased, development of flooding tends to be a continuous phenomenon wherein liquid droplets start to get entrained in the upward vapor flow. As the relative vapor velocity continues to increase and the column continues to approach a flooded state, the bulk density of the fluid within the packing void space increases until it approaches the condition depicted in Figure 3.

The condition depicted in Figures 3 and 4B is sometimes referred to as "Runaway Flooding". In Runaway Flooding, the liquid level continues to rise until the entire column is filled with frothy liquid. If Runaway Flooding is not stopped, the column can actually overflow and send liquid into the distillate process lines.

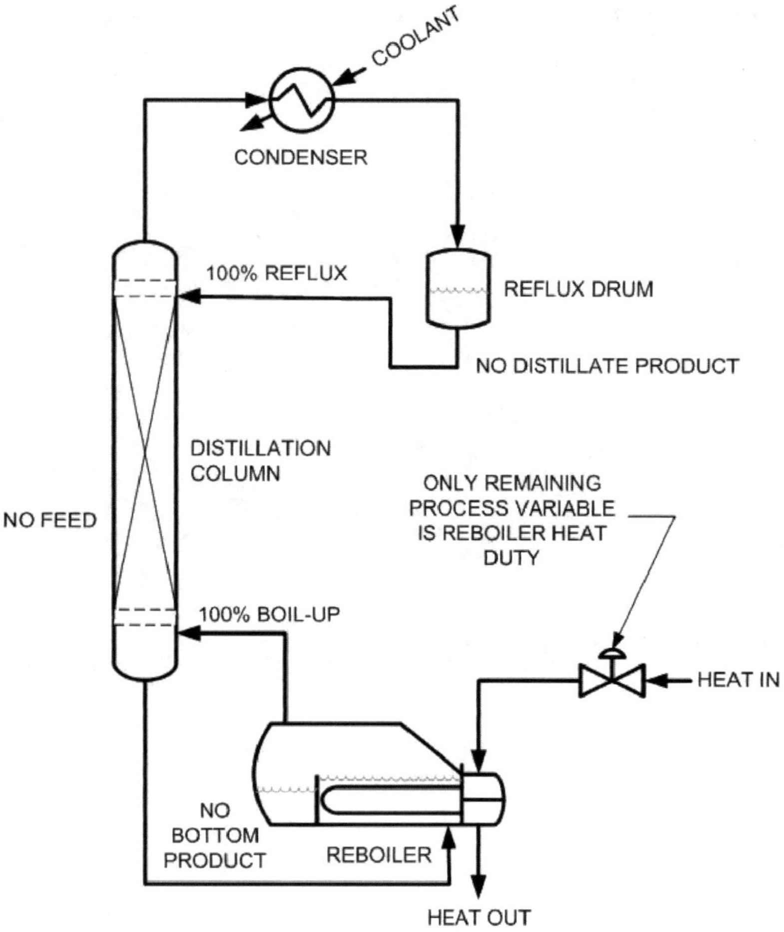

Figure 5 - 100% Reflux Mode

For any given distillation column, exactly what stage in the development of flooding can be detected will be dependent upon such factors as:

1. Column length over which the DP measurement is taken. Shorter column length will enable more sensitivity to detect flooding over that length of column. In actual practice, the DP across the stripping section and the rectification section of the column might be measured separately to detect flooding independently in each section.

2. [Differential Pressure] / [Absolute Pressure] ratio in the column . A greater [DP/AP] ratio will make the measurement more sensitive to detect flooding.

3. Presence and length of DP transmitter impulse lines. Long impulse lines tend to attenuate the high frequency signals this technique depends on to detect flooding conditions.

Ideally, one would like to detect flooding in the distillation column before it reaches the point of Runaway Flooding. Thus, an ideal diagnostic for distillation column flooding detection would show the column is approaching the Runaway Flooding condition. (i.e., one that shows the column is beginning to flood) This condition will be referred to as "Incipient Flooding."

Field Experiment

A field experiment was conducted at the J.J. Pickle Research Campus of the University of Texas at Austin, TX on March 5 - 6, 2008. The field experiment used an 18 inch inside diameter distillation column operated by the U.T. Separation Research Program. During this experiment, the column was operated in 100% reflux mode. In this mode, the reboiler's heat duty could be increased incrementally until flooding was observed. The distillation column was instrumented with Foundation Fieldbus devices, and DeltaV was used as the distributed control system. Figure 6 illustrates the operator interface for this distillation column.

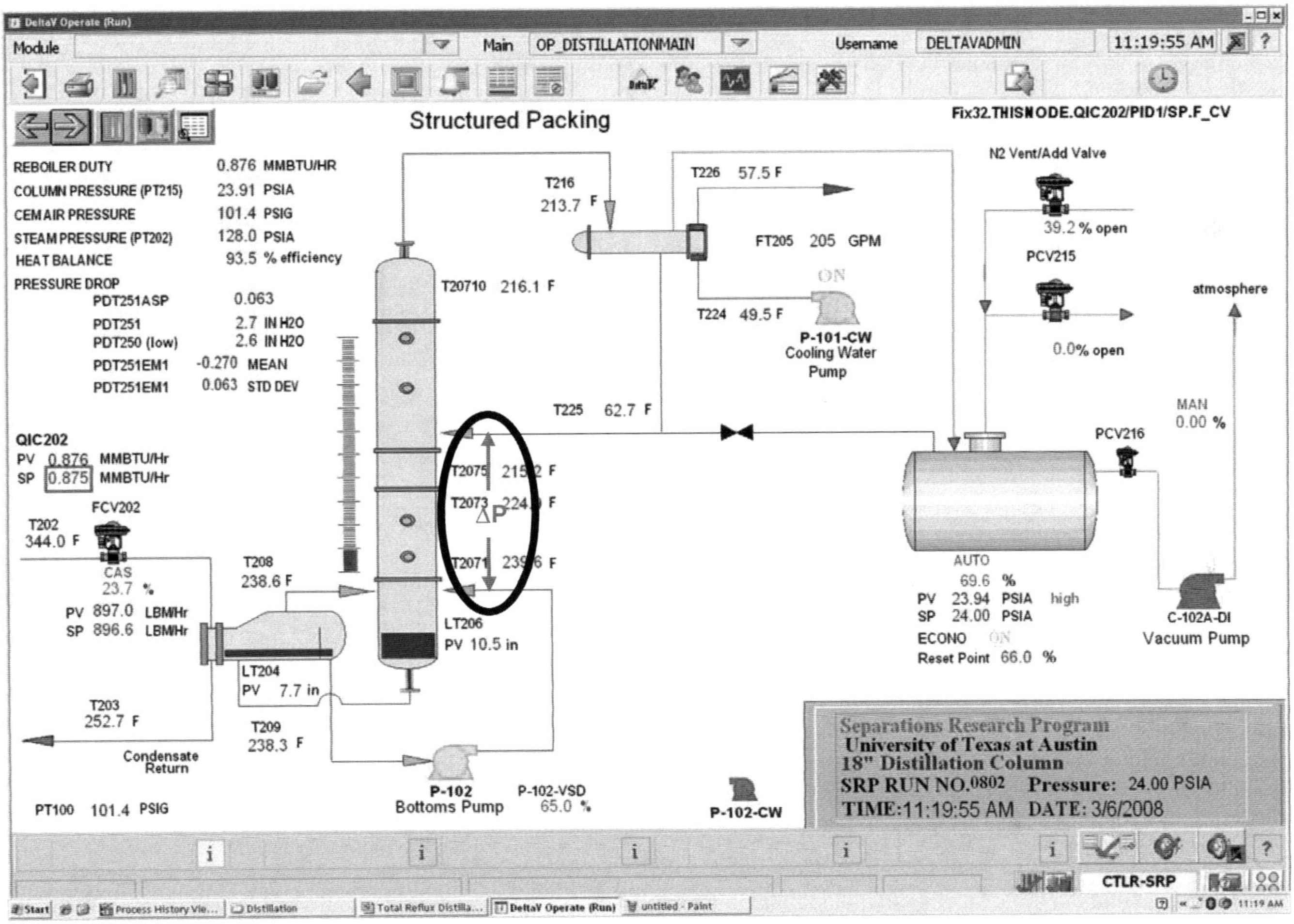

Figure 6 – Operator Interface for Distillation Column Control & Showing Location of DP Measurement

In this experiment, the structured packing only filled the bottom half of the column. A Rosemount 3051S DP transmitter (Range 1: -25 to 25 inches of water) with the Advanced Diagnostics option was installed on the distillation colu

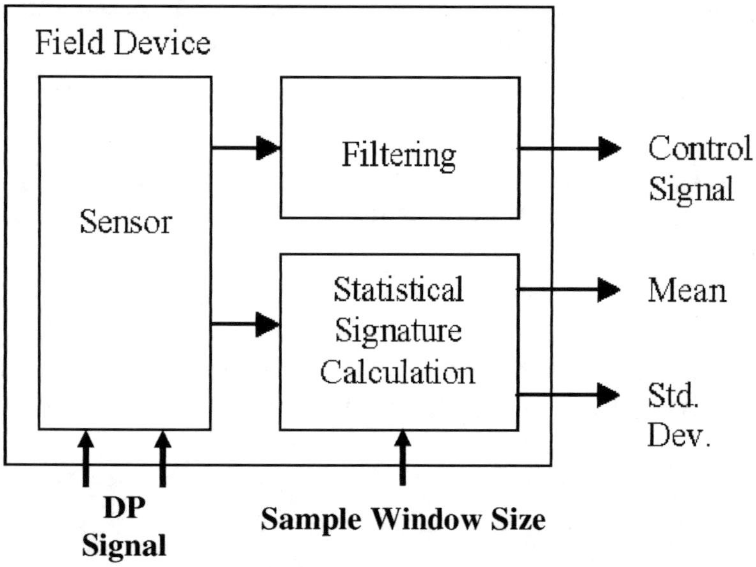

Figure 7 3051S Schematic Block Diagram

Figure 7 is a schematic block diagram of the transmitter used for this test. The sensor actually samples the DP signal at 22 samples/sec (22 hz). An on-board microprocessor, inside the transmitter, takes this raw 22 hz sensor signal and generates the conventional, low pass filtered, 4 – 20 ma control signal. The microprocessor also takes the same raw 22 hz signal, applies user-selectable filtering, and then calculates a trailing average mean and standard deviation. The same sample window size is used for calculating the trailing average mean and standard deviation. The user-selectable sample window size sets a trailing average time constant. This is effectively a low-pass or damping filter on the output from the statistical signature calculations. A specialized data acquisition system was used to log the pressure signal at the same rate as sampled within the transmitter, approximately 22 hz. By logging the raw sampled process data, it was possible to determine optimal configuration settings (e.g. filter and sample window size) for detecting flooding.

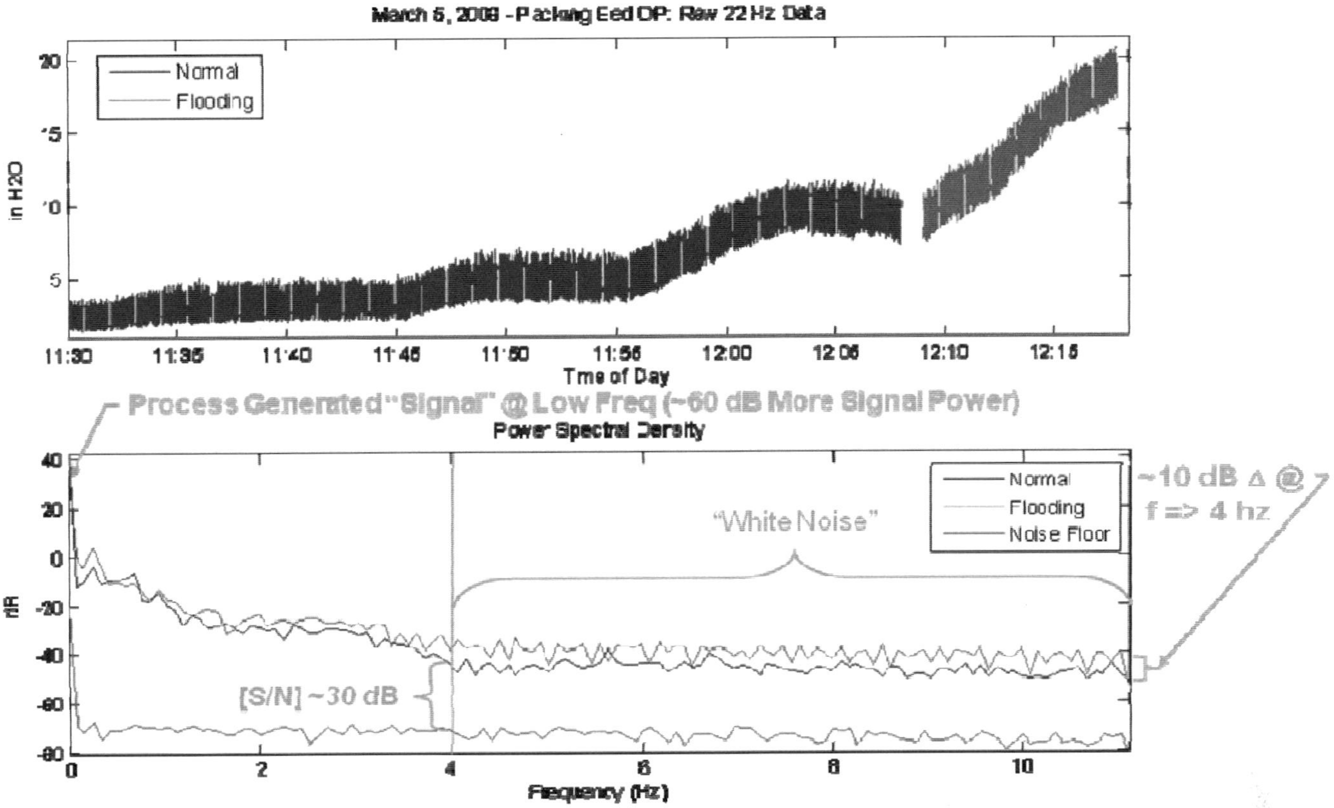

Figure 8 – Power Spectral Density Analysis, Flooding Experiment 1

A power spectral density analysis was done off-line in order to determine an optimal digital filter for this application. The top plot of Figure 8 shows the raw 22 hz data collected from the 3051S. This plot is divided into two segments: Normal and Flooding. The bottom plot in Figure 8 shows a power spectral density analysis on each of these data segments plus the signal generated by the transmitter when it was isolated from the process (Noise Floor). Because the signal was sampled at 22 hz, the range of the frequency spectrum analysis is from 0 to 11 hz, the so-called Nyquist frequency.

In the low frequency range (<= 2 hz), the power spectra between these data sets are fairly closely matched and overlap to a large degree. However, in the higher frequency ranges (=> 4 hz), there is approximately 10 db more signal power between the Flooding condition vs. the Normal condition. It is believed this signal power difference is due to the previously discussed phase inversion associated with flooding. Because the greatest difference in the frequency spectrum is at the higher frequencies, a good filter for detecting flooding would be one that attenuates lower frequencies, while passing higher frequencies. (high-pass filter) There is also about 30 db of difference between the transmitter's Noise Floor and the high frequency signals of interest. This [S/N] ratio of 30 db is important because it would be difficult to distinguish transmitter noise from the high frequency signals of interest.

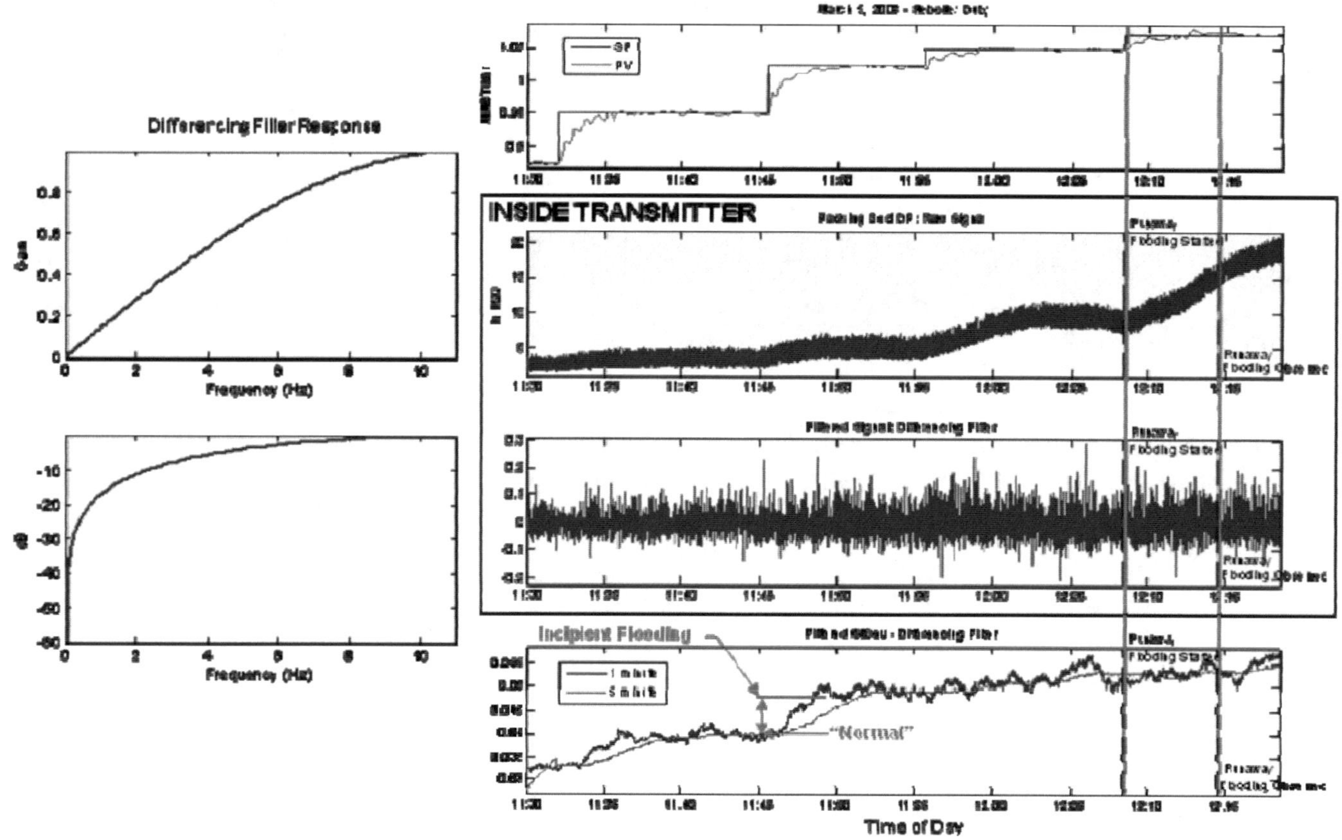

Figure 9 – Detecting Incipient Flooding Using The Differencing Filter

The advanced diagnostics option incorporates a user selectable (On/Off), first-order, digital high-pass filter (differencing filter). This filter cuts off the lower frequency signals from the sensor prior to the statistical signature calculations, while passing the higher frequency signals. The two plots on the left side of Figure 9 illustrate the frequency response of the differencing filter. The four plots on the right side of Figure 9 illustrate the triggering, signal processing, and detection of Incipient Flooding, all overlaid on the same time scale. The two plots inside of the blue box in Figure 9 graphically illustrate the signal processing inside of the transmitter.

The bottom plot of Figure 9 shows the standard deviation, using 1 and 5 minute sampling windows, of the high-pass filtered DP signal. Each time the reboiler heat duty was increased, the standard deviation also increased. In the Incipient Flooding condition, the standard deviation was significantly greater than in the Normal condition. Note: Because a 5 minute rolling window was used for the standard deviation calculation, it took at least that long for the standard deviation to reach its final value after the reboiler heat duty was increased. Note also: The standard deviation was even greater when the distillation column transitioned to the Runaway Flooding condition.

The filtered standard deviation clearly gives an indication that the distillation column is approaching the flooded condition. By trending this filtered standard deviation, the plant operator can be given an early warning that the distillation column is approaching the flooding point. In this way, an abnormal situation may be prevented.

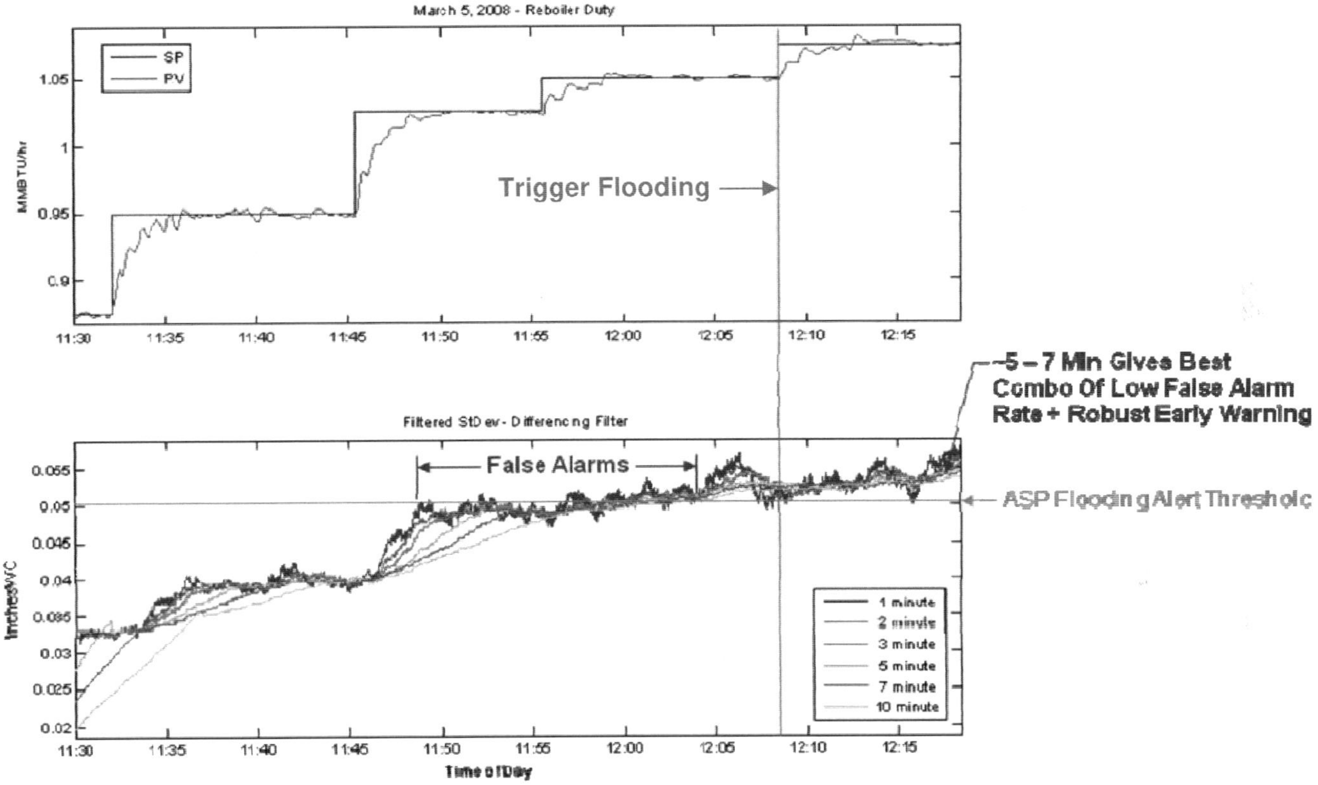

Figure 10 – Tuning The Sampling Window Time Constant

It should be noted that a variety of other high-pass filters were also tried, and using these different filters, both the filtered signal and the filtered standard deviation were plotted. However, none of them worked any better than the differencing filter for detecting incipient flooding.

The standard deviation of the signal output from a high-pass filter is intrinsically subject to high frequency "noise" due to process transients, setpoint changes, load changes, etc. Passing this signal through a low-pass filter will make the incipient flooding alert more robust against false alarms. The sampling window time constant for this filter must be tuned. Figure

10 illustrates the response to various time constants from 1 to 10 minutes. For this distillation column, time constants of 5 to 7 minutes gave the best combination of a low false alarm rate and relatively fast response time to an actual Incipient Flooding condition. A fast response time is important in order to be able to give the operator the earliest possible warning to correct the Incipient Flooding condition before it becomes Runaway Flooding.

Figure 11 illustrates another flooding test, in which the reboiler duty was repeatedly changed between the Runaway Flooding condition (reboiler heat duty = 1.05 - 1.07 MMBtu/hr) and the Normal condition (reboiler heat duty = 0.95 MMBtu/hr). The top plot shows the reboiler heat duty, while the bottom plot shows the 5-minute filtered standard deviation. Notice that each time after the process was changed from the Normal condition to the Runaway Flooding condition, the filtered standard deviation increased. Furthermore, notice that it was possible to

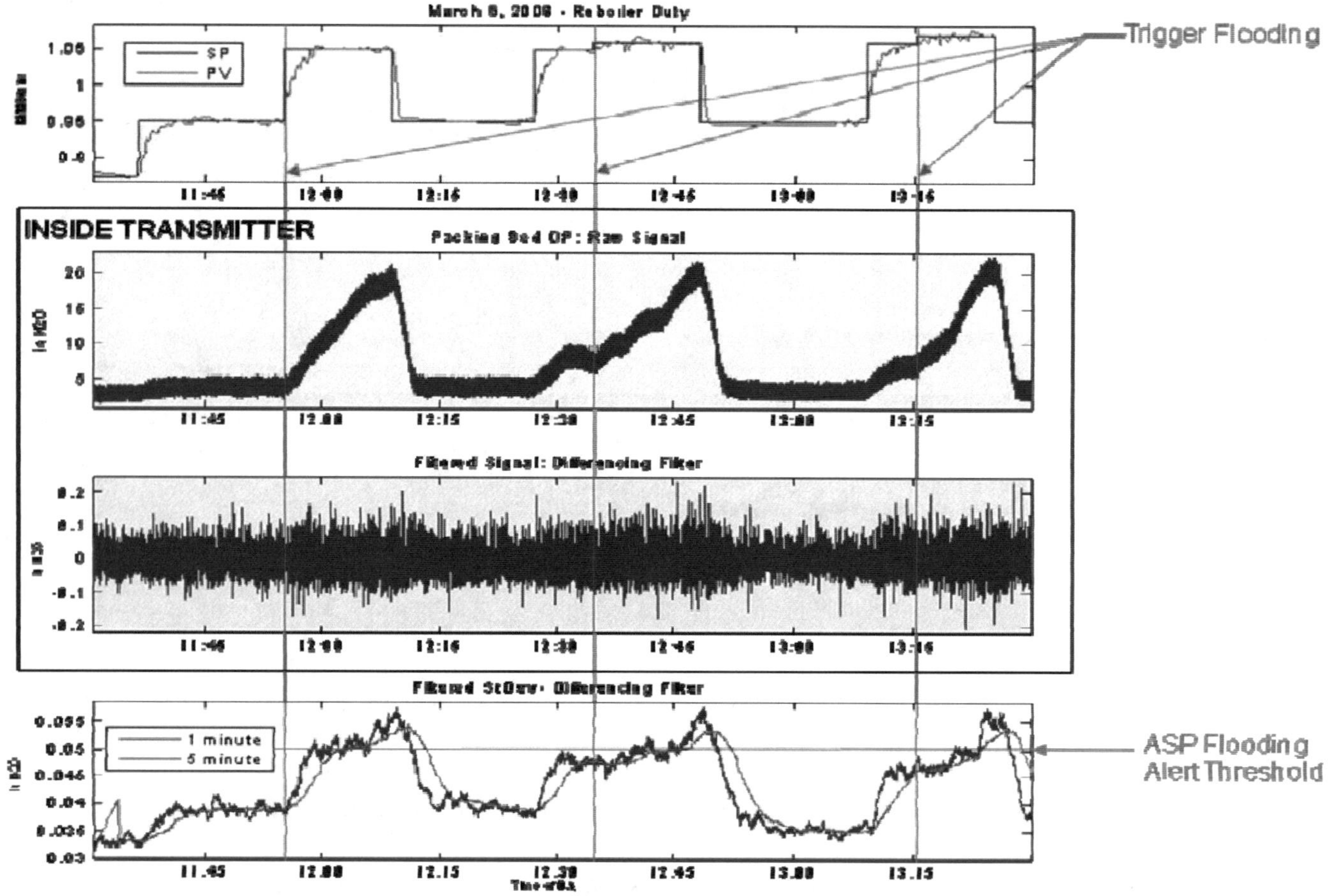

Figure 11 – Reproducible Flooding Alert

define a threshold on the standard deviation which detected Incipient Flooding in all three cases without any false alarms. When the standard deviation exceeded the threshold, it was a reliable indication that, although the separation quality is currently good, the distillation column was approaching the Runaway Flooding condition. This test demonstrated the repeatability of creating an incipient flooding alert based upon the filtered standard deviation.

Because this technique detects a phenomenon intrinsically associated with the onset of flooding, it should not be affected by other process changes such as fouling, internal damage, feedstock composition, etc. Whatever the root cause, whenever the distillation column transitions to Incipient Flooding, this technique should detect it.

Conclusion

Distillation Column Flooding is a phenomenon that can cause loss of separation and negatively impact the throughput and energy efficiency of a distillation column. The onset of distillation column flooding is associated with a change in the flow regimes of the gas and liquids flowing inside the column. The flow regime associated with flooding generates more high frequency white noise which can be detected in the DP signal across the column. This paper has discussed how to use this phenomenon to detect Incipient Flooding and then generate a robust Incipient Flooding alert . When an operator is made aware in real time that the column is approaching flooding, adjustments can be made to prevent the column from becoming completed flooded.

References

1. Nicholas P. Cheremisinoff, Ph.D., p. 163, Handbook of Chemical Processing Equipment, Butterworth-Heinemann, Woburn, MA, (2000)

2. ibid, p. 166

3. Norman P. Liebermann, pp. 256 – 280, Troubleshooting Process Operations, 3rd Edition, Pennwell Publishing Co., Tulsa, OK, (1991)

Development of a Renewable Energy-Based Solution for Saline Waters Desalination

Raafat Alnaizy[*] and Ahmed A Aidan
Chemical Engineering, American University of Sharjah, United Arab Emirates

2010 AIChE Spring Meeting
Grand Hyatt San Antonio San Antonio, TX
March 21-25, 2010

Abstract

It is vital to identify and develop diverse and sustainable energy alternatives to overcome the current and future energy shortage. Energy costs can be greatly reduced by utilizing a natural and "free" source of energy. In this work, we present the findings of using renewable energy sources (solar) as a potential solution to meet the potable water needs. The project addresses these issues by examining a multi-effect, multi-stage (MEMS) flash distillation system for desalination using solar heating and the sensible heat that can be extracted from a Salinity Gradient Solar Pond (SGSP). The project involved a comprehensive testing and analysis over a broad range of operational conditions, with the thermal energy derived from a SGSP. The relationship between system performance and operating conditions was analyzed to determine the economic and technical viability of the system. Mathematical and computer models were developed to describe the physical behavior of the system and to predict its performance. The proposed desalination process is expected to meet all imposed criteria and conditions, i.e. zero discharge with a very high recovery ratio of 82%. In addition, all the thermal energy needed for the MEMS, which represents more than 90% of the total energy, will be derived from a solar pond. The close-to-saturation brine will be sent to a shallow man-made pond to evaporate all the water and collect the remaining salts as crystals and guarantee the zero-liquid discharge process. The process is far less costly than the other processes, as the site conditions are favorable for salinity-gradient solar ponds.

Keywords: desalination, Solar Pond, renewable energy, multi-effect Evaporator
*Author to whom correspondence should be addressed P.O. Box 26666, Sharjah, UAE
Phone: (971) 6-515-2710, Fax: (971) 6-515-2979, Email: ralnaizy@aus.edu

Introduction

The main source of potable water in the arid United Arab Emirates (UAE) is the desalination of seawater and the brackish or hypersaline groundwater found along the Arabian Gulf coastline. The Sharjah Electricity and Water Authority (SEWA) currently produces potable water and electricity by using a variety of fossil fuels, including natural gas and diesel to desalinate seawater and/or brackish groundwater. Due to affects on environmental, sustainability, foreseeable energy scarcity, etc., there is a pressing need for sustainable solutions in the near future (1). The desalination of sea or brackish water offers an alternative solution where the availability of water cannot be increased by using conventional resources or by recycling methods. Desalination is becoming a solution for water scarcity in most arid countries such as the Arabian Gulf states, which have more than 50% of the total installed plant desalination capacity and other small islands all over the world. New technological advances in the last 30 years tremendously reduced the capital cost and the energy consumption so that desalination projects allowed for alternative solutions to water development (2). However, desalination projects are still not very affordable to be easily accommodated by the economies of many countries, energy consumption is still comparatively high. Distillation units routinely use designs that conserve as much thermal energy as possible by interchanging the heat of condensation and heat of vaporization within the units. The major energy requirement in the distillation process thus becomes providing the heat for vaporization to the hypersaline groundwater.

Scope and Objectives

Groundwater of high Total Dissolved Solids (TDS) in the range of 82,000 ppm (hypersaline water) is found in many towns and cities in the UAE. From an environmental point of view, for one of the locations, it was determined that desalination of the available hypersaline groundwater for human needs may be preferable to using the gulf waters. During the same initiative, there was a need to explore options of recycling and reusing greywater after thorough treatment. The overall goal was to produce potable water from the available hypersaline water with zero-liquid discharge. The design would also meet a low energy consumption criterion comparable to best practice desalination (< 5 kWh/m^3), and meets drinking water quality standards of Regulation Supervision Bureau (RSB) and US Environmental Protection Agency (USEPA). As it is very difficult to desalinate water of such high TDS, the scope of work was to develop a method and process that will result in an efficient and cost effective system. We considered all potential sustainable solutions to bring the best technology to produce potable water with initial water capacity of 1,000 m^3/day from the hypersaline groundwater sources combined with the recycled locally generated greywater under the constraint mentioned above.

Process Description

To achieve the required outcome under the given constraints, considering the high levels of salinity of the hypersaline groundwater, extensive research was conducted to identify a method that would result in an efficient and cost effective system. The credibility of the approach is critical to establishing a successful design for the greywater, and the hypersaline groundwater. It was identified that the solution lay in the reduction of the salinity of the hypersaline groundwater. It was hypothesized to dilute the hypersaline groundwater with treated greywater, which would be generated and treated insitu by the use of a membrane method such as reverse osmosis, nanofiltration, or ultrafiltration. Disinfection treatment by chlorine or UV irradiation is not required since the waters will be further treated. The TDS of the hypersaline groundwater decreased to about 50,000 ppm using a mixing ratio of two to one (saline water to greywater). To desalt the resultant stream, MEMS flash in combination with a closed-cycle salt-gradient solar pond was determined to be the most suitable desalination layout. It has a very high recovery ratio of 82%. In addition, all the thermal energy needed for the MEMS, which represents more than 90% of the total energy, will be derived from a solar pond. The nearly saturated brine is sent to a shallow man-made pond to evaporate all the water and collect the remaining salts as crystals and guarantee the zero-liquid discharge process. This process is far less costly than the other processes studied, as the site conditions are favorable for salinity-gradient solar ponds. Combining the solar pond technology with MEMS flash is expected to meet all imposed criteria and conditions, i.e. zero discharge.

Results and Discussion

The proposed process was designed and simulated to have a minimum environmental impact and low capital and operational costs. The salinity of the groundwater was reduced by dilution with greywater. Heat in the UAE reaches an average of 43°C in summer; at this temperature, the bottom of the pond can easly reach 90 °C. A large solar pond averages the energy it receives from the sun over about six weeks (3). Pipes carrying brine are laid at the bottom of the pond; this allows the heating of the brine inlet to the MEMS. Table 1 shows the average monthly temperatures in the UAE as reported by UAE University.

Table 1: Average Monthly Temperature in the UAE

	Jan	Feb	Mar	April	May	June	July	Aug	Sept	Oct	Nov	Dec
Temperature in °C	24	25	29	33	38	39	43	43	39	35	30	26

To make the plant more economical and attractive, the processes were mutually integrated. The energy consumption was lowered as more heat is recovered. The heat exchange between the processes can reduce waste heat, CO_2 and SO_2 emissions, as well as pollution. The design method focused on saving energy. Energy integration between several streams and units were applied to give more cost-effective operations. The entire process was simulated on Super Pro designer. The overall flow diagram of the process is shown in Figure 1. It was assumed that each flash recovery is between 10% and 50% of the hypersaline groundwater. The vacuum on each stage is in agreement with industry best practices. The proposed process design has a robust heat recovery and integration. The process has a high recovery of 82% and more than 90% of the required energy is derived from solar energy using the solar pond.

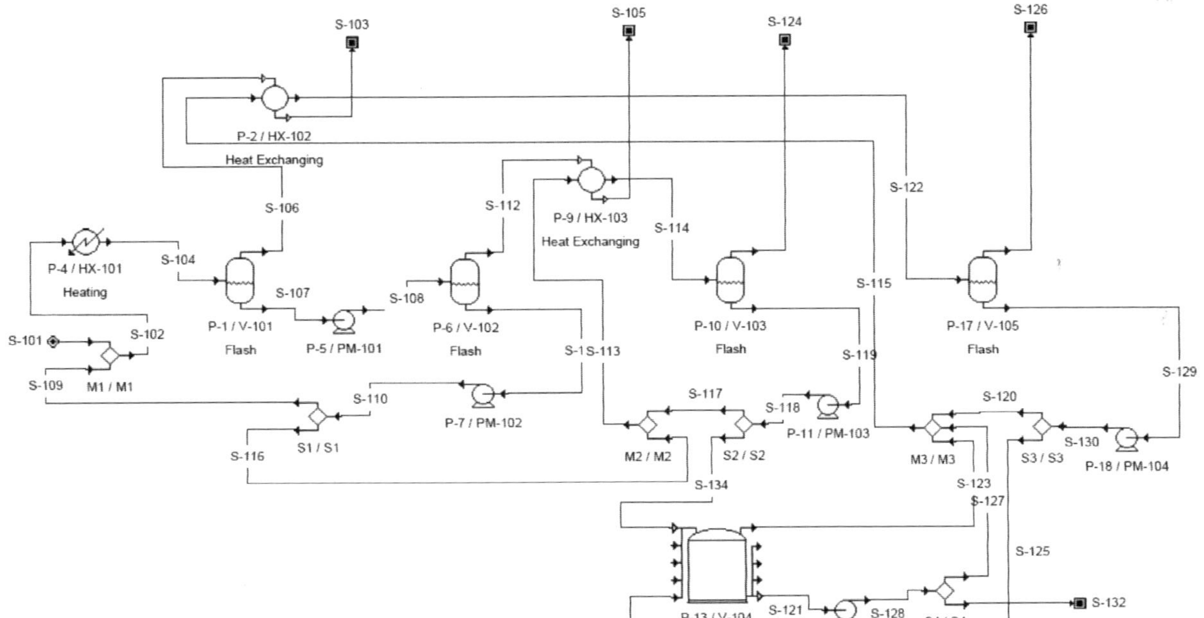

Figure 1: The proposed process flow diagram for the desalination of hypersaline groundwater

The hypersaline groundwater is first mixed with the treated greywater. Stream S101 contains the diluted groundwater with Sodium Chloride (NaCl) loading of 5.2 % (assuming all dissolved solids to be NaCl). The mass flowrate of stream S101 is 59,000 kg/h. It enters the process under atmospheric pressure and 32 °C. It is mixed with the recycle stream S109 from the second flash drum stream. Stream S109 has a mass flow rate of 1,030,000 kg/h at 72 °C. The resultant stream S102 (8.1% NaCl) flows onto the solar pond with a mass flowrate of 1,090,000 kg/h at 70 °C. The solar pond is depicted as heat exchanger HX101 in the simulation, Figure 1. After exchanging heat with the solar pond, the temperature of stream S104 rises to 85 °C; it flows onto the first flash V101 where the pressure decreases from atmospheric pressure to 43 kPa. The water is flashed at the saturation temperature of 78 °C. The vapor stream (S106) flowing at a rate of 12,600 kg/h is cooled by exchanging heat with the feed to the last (fourth) flash vessel that is operating at the lowest vacuum and lowest temperature. This heat exchange is taking place at HX102 (Figure 1). It allows for higher evaporation rate due to the large temperature difference between the first and last vessel. The brine in the bottom stream (S107) of the first flash (V101) flows onto the second flash drum (V102)) operating at a lower pressure (34 kPa). The procedure is repeated onto the third and fourth flash vessels as shown in Figure 1. Stream S110 is split into two streams (S109 and S116). The splitting ratio is 96.5% for stream S109. The recycling ratio is very high because the liquid must circulate in the stage in the maximum numbers of cycles as possible to ensure maximum vaporization. These two streams have an identical composition of 8.3% NaCl. Stream S109 is the recycling stream mentioned above that was mixed with the diluted feed. Stream S116 is mixed with the recycled stream of the bottom of the third flash (S117). The mixture is then heated in heat exchanger HX103 by exchanging heat with the vapor product of the second flash. Tank V104 is a reconcentrator to maximize the circulation of the liquid, to ensure a uniform solution thus minimizing the size of the fourth flash. The tank has an additional inlet S125 that is part of the outlet of the fourth flash vessel. Furthermore, this tank has two outlets: S121 and S123. The flow rate of S123 is usually equal to zero nevertheless it was designed as a vent for safety precaution. There may be emissions from the tank or the tank may overflow. Stream S121 is pumped to a splitter where it is split into the final brine stream S132 (5%) and stream S127 (95%) that joins other streams (S120 and S123) to form stream S115. Stream S115 is passed into heat exchanger HX102 for heat recovery from the outlet vapor of the first flash vessel. It is fed to the fourth flash vessel as stream S122. Streams S103, S105, S124, and S126 are the distillate products of the flash vessels, these are all connected to the suction header from the eductor-jet pump used to create vacuum.

Conclusions

The study proposes a system for the desalination of hypersaline groundwater (in excess of 80,000 ppm) diluted with treated greywater. The greywater was treated in a separate process using sand filters, activated carbon filters, and ultra filtration. The salinity of the groundwater was reduced to reasonable levels (around 50,000 ppm) then treated using a four-effect multi-stage flash distillation system. The recovery of the MEMS unit is high due to certain recycle ratios and heat integration. The heat integration/exchange between the first and last flash yields a very high recovery/production in the last vessel due to the high temperature difference between the first flash vapor and the inlet to the fourth flash. A salinity-gradient solar pond was used as a source of heat for the proposed process. The salinity of the gradient solar pond was maintained using an evaporation pond that was continuously providing concentrated brine to the lower zone to compensate for the salt that had naturally diffused to the surface. The salt that drifted to the surface was then recycled. A submerged mesh of pipes within the pond containing the flowing water was utilized because of its simplicity. The temperature in the storage zone was 95 ^{o}C. The proposed process to desalinate the hypersaline groundwater meets all constraints; low energy consumption, high recovery, zero-liquid discharge, and pure water (~4 ppm TDS).

References

1. Glueckstern, P., 1995, "Potential Uses of Solar Energy for Seawater Desalination, Desalination", 101 (1995): 11-20.
2. Al-Shammiri, M. and Safar, M., 1999, "Multi-Effect Distillation Plants: State of the Art," Desalination, 126 (1999): 45-59.
3. Li, Luming, 1992, "Understanding of Design Features and Performance Characteristics of a Solar Pond-coupled Multistage Flash Desalination System", Master's Thesis, Mechanical Engineering Department, University of Texas at El Paso.

Simulation of Batch Reactive Distillation for Methyl Acetate Production

Prof. Nada B. Nakkash[1] **Dr. Naseer Al-Habobi**[1] **Haider M. Hamodee**

[1]*Dept. Chemical Engineering, College of Engineering,*
Al-Nahrain University, Baghdad - Iraq.
E-mail: nadanakkash@yahoo.com

Keyword: *Reactive Distillation, Estrification, Batch Reactive Distillation, Simulation, Multicomponent Distillation*

Abstract

The present work is concerned with simulating a batch reactive multistage multicomponent distillation. Rigorous model was used to build the simulation program using $MatLab_{6.5}$ to solve MESHR equations. The validity and the accuracy of the developed program model were checked with previous work for the estrification of acetic acid with methanol to produce methyl acetate and water using sulfuric acid as a homogeneous catalyst.

The influence of various parameters, such as number of stages, batch time and liquid hold up on the performance of batch reactive multicomponent distillation was shown through the vapor and liquid profile, temperature, reaction rate and liquid composition profiles.

The results show that the maximum recovery obtained for methyl acetate was 90.7% and the overall conversion to methyl acetate was 49.41% under best operating conditions of 1.7hr batch time, 6 theoretical stages, 2.5 reflux ratio, 1atm pressure,310.15K feed temperature with constant liquid hold up through out the tower.

Changing the liquid hold up throughout the tower will affect the performance of tower and increases the overall conversion of methyl acetate to 50.4% and the recovery of methyl acetate increased to 92.5%.

Introduction:

Reactive distillation (RD) combines chemical reaction and distillation into a single process unit. The chemical reaction usually takes place in liquid phase or at the surface of a catalyst in contact with the liquid phase.

Reactive distillation has proved to be an important process alternative to the conventional reactor separator configuration. Distillation with chemical reaction also offers advantages in overcoming the equilibrium limitation for reversible reactions and separations of undesired byproducts. In a typical BRD process, reaction and separation occur simultaneously. The removal of one or more products from the reaction mixture not only increase the conversion of equilibrium limited reactions but also increases or decreases the reaction vessel temperature if the product removed is lower or higher boiling, respectively (1).

Kian (2) discussed that batch distillation which is often directly coupled with chemical reactors such that the reactor acts as the reboiler of the column. This combined process increases the conversion of equilibrium reaction, making high conversions easily achievable, by the removal of one of the products through the distillation column.

Sundmacher and Steyer (3) stated that the crude lactic mixture can be reacted with alcohols like methanol in a batch reactive distillation apparatus with simultaneous distillation of the voltaic methyl lactate as a distillate. Methyl lactate is fed to another batch reactive distillation apparatus where hydrolysis of the methyl lactate to form lactic acid and methanol is performed.

Mujtaba and Macchietto (4) considered a theoretical comparative study with conventional and unconventional columns for simultaneous chemical reaction and distillation, with a reversible reaction.

$$A + B \leftrightarrow C + D$$

The reaction products are C and D, with D being the most volatile component and C being the least volatile component in the reaction mixture. The products are removed simultaneously with good conversion and the amount of the distillate product is achieved.

The basic equations for the batch distillation calculations are the same as those for continuous distillation processes. Material balance, phase equilibrium correlations and energy balances are formulated for simulation purposes. The balance equations are applied on each tray.

BatchFrac is a batch distillation model that solves unsteady-state heat and material balance equations by rigorous equations. These equations describe the behavior of a multistage batch distillation column. BatchFrac applies rigorous heat balances, material balances, and equilibrium relationships at each stage. BatchFrac calculates the profiles of column composition, temperature, pressure, and vapor and liquid flows as a function of time (5).

Huckaba and Danly (6) developed a computer program that assumed constant mass tray holdups, adiabatic tray operation, and linear enthalpy relationships included energy balances around each tray.

Meadows (7) developed the first rigorous multi-component batch distillation model, based on assumptions of equilibrium stages, perfect mixing of liquid and vapor phases at each stage, negligible vapor holdup and constant molar holdup.

Distefano (8) extended the model and developed a computer-based-solution procedure that was used to simulate successfully several commercial batch distillation columns.

Boston (9) extended the model, provided a variety of practical sets of specifications, and utilized modern numerical procedures and equation formulations to handle efficiently the nonlinear and often stiff nature of multi-component batch distillation problem.

Marta et.al. (10) described a case study to fit a rigorous model using the commercial package HYSYS, for multi-component batch distillation processes.

Galindez H. and Fredenslund (11) developed to simulate the operation of multistage, multi-component batch distillation operations using rigorous algorithms for the integration of the differential equations.

Ping (12) formulated a rigorous model for the dynamic simulation of a multiphase batch distillation in the software gPROMS.

Jimenez et. al. (13) used rigorous models as an alternative to predict the concentration profile and to specify the optimal switching time from products to slop cuts for nonlinear dynamic modeling of multi-component batch distillation.

Mujtaba and Macchietto (14) described the dynamic optimization of batch reactive distillation with an objective, so they used a detailed model including mass and energy balances.

Lee et al (15) studied a batch reactive distillation with two feed to produce pure ethyl acetate for the stoichiometric feed of acetic acid and ethanol. They concluded that this type cannot produce pure ethyl acetate.

Elgue (16) developed the rigorous model to solve the differential equations that consist of total and component material balances, total energy balance, negligible vapor holdup, and constant liquid holdup for batch reactive distillation.

Hamodi (17) constructed a simulation model to analyze multi component multistage batch reactive distillation producing methyl acetate using Mat lab program.

Patel et al (18) derived a mathematical model and simulation for ethyl acetate synthesis. The differential algebraic equations solved using forth order Runge-Kutta method in Mat lab program.

Present Work:

The analysis emphasized to simulate multistage unsteady state batch distillation with of N feed components into relatively pure products.

The stage temperature, vapor and liquid flow rates, liquid holdup, reaction rate, and composition for all components with time are required and are calculated by solving (MESHR) equations for batch with chemical reaction for all trays.

The simulation model based on the multi-component batch rectification operation similarly to a conventional batch distillation column except that for the reaction occurs initially in the reboiler. This system can be described as a "batch reactor-rectifier" Figure1. It consists of a reaction vessel (reboiler) and a fractionating column on top. The vapors coming from the topmost plate of the column are condensed in the total condenser, and the distillate is collected in the receiver, where methanol reacted with acetic acid with sulfuric acid as homogenous catalyst in reaction vessel (reboiler) to produce water and methyl-acetate. All components are vaporized, so that the reaction takes place in all trays, the desired product input to the condenser section, and collected in the receiver and produced as top product. The rate of reaction is a function of temperature, catalyst concentration of the products and reactants. In the present work the following assumptions were made for the developed model:

a- The liquid-phase reaction takes place in all stages.
b- The separation system has sufficient stages and reflux to achieve the desired distillate composition.
c- There are no reactive or non-reactive azeotropes in the system.
d- A total condenser produces a saturated liquid product.
e- Constant plate efficiency.
f- Neglect the vapor holdup.
g- Perfect mixing in all trays.
h- Constant pressure drop.
i- Variable liquid holdup.
j- Ideal vapor phase and non-ideal liquid phase.
k- Homogeneous catalyst (sulfuric acid).
l-

The estrification reaction is:

$$CH_3COOH + CH_3OH \rightarrow CH_3COOCH_3 + H_2O$$

Feed system and compositions studied are given in Table (1). All physical and chemical properties are taken form literature (17). Table (2) gives the initial input data (16, 17) to the developed simulation model computer program to check its validity.

The variables studied are batch time, number of theoretical stages and liquid holdup.

The material and energy balances were carried out for each component using rigorous method plate to plate calculation for unsteady state and with chemical reaction. In all cases the feed is assumed as liquid at its boiling point.

The composition, flow rates of vapor and liquid, reaction rate, temperature profiles and liquid holdup for each stage were obtained using developed model.

Mathematical Model:

Theoretical model for an equilibrium stage consider a batch unsteady-state reactive distillation column consisting of a number of stages arranged in a counter current cascade Figure (1) and a general schematic representation of equilibrium stage j Figure (2), where the stages are numbered from top to bottom with the condenser as a first stage and the reboiler as the nth stage. The model equations are:

-The overall material balance equations (M-equation).

$$V_1 y_{i,1} - L_o x_{i,o} - D x_{i,D} = \frac{d(M_o x_{i,o})}{dt} \quad (1)$$

- Mole fraction summations equations (S-equations).

$$(Sx)_i = \sum_{i=1}^{m} y_{ij} / K_{i,j} - 1 = 0 \quad (2)$$

$$(Sy)_i = \sum_{i=1}^{m} x_{ij} * K_{i,j} - 1 = 0 \quad (3)$$

-Phase equilibrium relations (E-equations).

$$E_{i,j} = y_{i,j} - K_{i,j} x_{i,j} \quad (4)$$

-The overall energy balance equations (H-equations)

$$V_1 h_{V1} - (L_o + D) h_{Li_o} = Q_{Nt+1} + \frac{d(Mc \cdot h_{Li_o})}{dt} \quad (5)$$

-Reaction rate equations (R-equation)

$$\sum_{j=1}^{m} \Delta R_j - \Delta R = 0 \quad (6)$$

The rate of reaction is a function of temperature composition (16, 17).

$$Rate(j) = \Delta R_j = Kester * \exp(-E/R * Ti(j))))$$
$$* (Cc_m(j) * Cc_a(j) - (Cc_w(j) * Cc_{ma}(j)) / Keq) \quad (7)$$

The solution of (MESHR) equations for multi-component batch distillation with chemical reaction is obtained by calculating a set of temperature, phase rate, batch time, rate of reaction and compositions which satisfies all the equations of the model that described above (i.e. (MESHR) equations).

The determination of phase composition and its temperature is calculated by the solution of material balance equations.

1. The overhead section:

$$dxi,o/_{dt} = -\left(\left(L_o + D + Mc_o \Big/ Mc_{i,o}\right)\right) * X_{i,o}$$
$$+ \left(\left(V_1 * K_{i,1}\right) \Big/ Mc_{i,o}\right) \cdot X_{i,1} + \Delta R_i + vol \quad (8)$$

2. Stage section:

$$dxi,j/_{dt} = \left(L_{j-1} \Big/ Mpi_j\right) * X_{i,j-1}$$
$$-\left(\left(L_j + (K_{i,j} * V_j) + dM_j/dt \Big/ Mp_{i,j}\right)\right) * X_{i,j} +$$
$$\left(K_{i,j+1} * V_{j+1} \Big/ Mp_{i,j}\right) * X_{i,j+1} + (\Delta R_i + vol) \quad (9)$$

3. Reboiler section:

$$dxi,o/_{dt} = \left(L(1) \Big/ Mr\right) * X_i(1)$$
$$-((((V(1) * K_i(1)) + Mt) / Mr) * X_i(1)) + \Delta R_i + \text{vol} \quad (10)$$

Where

$$Mr_j = M^o_{Nt+1} - \sum_{j=0}^{Nt} M_j - \int_0^t Ddt \quad (11)$$

The material balance equations are reduced to a tri-diagonal matrix form.

$$\begin{pmatrix} B_1 & C_1 & 0 & 0 & 0 \\ A_2 & B_2 & C_2 & 0 & 0 \\ 0 & A_j & B_j & C_j & 0 \\ 0 & 0 & A_{j-1} & B_{j-1} & C_{j-1} \\ 0 & 0 & 0 & A_{Nt} & B_{Nt} \end{pmatrix} \begin{pmatrix} X_{i1} \\ X_{i2} \\ X_{i,j} \\ X_{ij-1} \\ X_{iNt} \end{pmatrix} = \begin{pmatrix} dxi,1/dt \\ dxi,2/dt \\ dxi,j/dt \\ dxi,j-1/dt \\ dxi,Nt/dt \end{pmatrix} \qquad (12)$$

This set of equations may be formally written as the following matrix equation:

$$A.X = dx_{i,j}/dt \qquad (13)$$

The mole fraction $x_{i,j}$ from the algorithm Matrix is calculated using eigen-value. The values of mole fraction $x_{i,j}$ are corrected to provide better values of the assumed iteration variables for the next trial; thus, for each iteration if the computed set of xi,j values for each stage is not satisfy the summation constraint given by equation 14. The values of xi,j can be normalized using the following relation:

$$(x_{i,j})_{normalized} = \frac{x_{ij}}{\sum_{i=1}^{m} x_{ij}} \qquad (14)$$

After normalization of $x_{i,j}$ the correct temperature $T_{i,j}$ is calculated for each component i by the same loop for vapor-liquid equilibrium for the calculation of partial pressure $P_{i,j}$ from Antoine equation, distribution constant Ki,j, and the bubble point calculation for xi,j after normalizing as in equations below:

$$P_{i,j} = \exp\left(A_i + B_i / C_i + T_i\right) \qquad (15)$$

$$K_{i,j} = \frac{y_{i,j}}{x_{i,j}} = \gamma_{i,j}\frac{p^o_{i,j}}{p} \qquad (16)$$

In the present work $\gamma_{i,j}$ is calculated using Unifac equation (17,19).

$$Sum(j) = \sum_{i=1}^{m}(K_{i,j}X_{i,j}) - 1 = 0 \qquad 1 \le j \le Nt \qquad (17)$$

The modified **H-equations** are obtained first by calculating the liquid phase enthalpy HL_i, and the enthalpy of vapor phase HV_a. Second calculating the vapor flow rates V_{j+1}, then the heat supplied to the condenser Q_c.

The vapor enthalpy is

$$HV_{ai}(j)=h_{oi}+C1iV*(T(j)-T0)+C2iV*((T(j)^2-T0^2)/2)+C3iV*((T(j)^3-T0^3)/3)+C4iV*((T(j)^4-T0^4)/4) \qquad (18)$$

The liquid enthalpy is

$$HLi_j = \sum_{i}^{Nt} X_{i,j}H_{Li} \qquad (19)$$

A new set of V_{j+1} tear variables are computed by using the following modified energy balance, which is obtained by the calculation of liquid and vapor enthalpies:

$$V_{j+1}=\left[\left(\frac{1}{(HVq_{j+1}-H_{Lij})}\right)\bullet\begin{pmatrix}V_j\cdot(HVq_j-H_{Lij})-\\ L_{j-1}\cdot(H_{Lij-1}-H_{Lij})+Mn_j\cdot\frac{dH_{Lij}}{dt}\end{pmatrix}\right] \quad (20)$$

Where:

$$\frac{dH_{Lij}}{dt}=H_{Li(i,j)}-H_{L(i,j-1)} \quad (21)$$

The liquid holdup for stage j is:

$$Mn_j=\frac{1000*Gp}{\sum_{i=1}^{N}x_{i,j}*Wm_i/Den_i} \quad 2\le j\le Nt-1 \quad (22)$$

The liquid holdup at stage $j=1$ and $j=N_t$ are:

$$Mc_j=\frac{1000*Gc}{\sum_{i=1}^{N}x_{i,j}*Wm_i/Den_i} \quad j=1 \quad (23)$$

$$Mt_j=Mr_j-(V_{j+1}-V_j)*Stime \quad j=Nt \quad (24)$$

Where

$$Mr_j=\frac{1000*V_{reboiler}}{\sum_{i=1}^{N}x_{i,j}*Wm_i/Den_i} \quad j=Nt \quad (25)$$

The heat supplied to the condenser Q_c, is calculated by equation below:

$$Qc=V_j*\left(\sum_{i}^{N}X_{i,j}(Hva_{i,j}-H_{Li,j})\right)-Mr*HLi_j \quad (26)$$

For non-ideal system heat of mixing was added to the total enthalpy liquid phase:

$$H_{mix}=RT\sum_{i=1}^{C}(x\ln\gamma_i) \quad (27)$$

$$HLi_j=\sum_{i}^{N}X_{i,j}H_{Li}+H_{mix} \quad (28)$$

The reaction term is added to the enthalpy in Liquid phase (17):

$$H_{Li}=\Delta Hr*rate(j)*Vol \quad (29)$$

Results and Discussion:

When the reaction is added to the model calculation, together, with the necessary modifications. The simulating program was run to solve the reaction of acetic acid and methanol with sulfuric acid as homogeneous catalyst to produce water and methyl-acetate with reaction corrections. The developed model accuracy and validity calculation was checked with the conditions and experimental data for the same system taken from previous work (16).

Also, the final results of the vapor and liquid flow rates, temperature, liquid holdup in (reboiler, plates and condenser), and liquid composition profiles, using the developed model computer programs $MATLAB_{6.5}$ for batch distillation with chemical reaction to produce water and methyl-acetate were compared with the results obtained by Elgue (16).

Figure 3 shows the changes in still liquid composition of each feed component with batch time. As batch time is increased, the concentrations of methyl acetate, methanol and acetic acid decreased while for water increased at the reboiler section.

Figure 4 shows the comparison of the liquid composition profile at the still of each component versus batch time for present work with the experimental data (16), so it gives a good agreement, with a small deviation due to thermodynamic models used, the non-ideal model and the kinetic model simplicity.

Figure 5, shows the composition profile of methyl acetate that increased form the reboiler to distillate section. The mole fraction of methyl acetate in the distillate is 0.247 %, the conversion is 49.41% and the recovery fraction is 0.907.

The effect of process variables such as number of stages, batch time and liquid holdup on the performance of multi-component batch reactive distillation was studied. All the results obtained for the total vapor and liquid flow rates, temperature, reaction rate and the liquid composition profiles at various process variables are presented graphically for comparison.

When the batch time changed to 1hr the mole fraction of the methyl-acetate at top decreased to 0.225 with 45%conversion and for 3hrs batch time the mole fraction of the methyl-acetate decreased to 0.217 with 43.4%conversion. From these one can conclude that changing the mole fraction of methyl acetate depends on the simulation curve of the desired product vs. batch time this curve increased until it reached to 1.7hr, this curve begin to decrease until it becomes constant at 5hrs batch time Figure 6.

The effect of changing number of stages on the reaction of acetic acid and methanol to produce methyl-acetate and water Figure 5 shows that the liquid composition of methyl-acetate at the top section using 6 theoretical stages is 0.247, with 49.4%conversion. When the number of stages changed to 4 theoretical stages the mole fraction of the methyl-acetate decreased to 0.05 at the top as shown in Figure 7 with 10% conversion. Increasing the number of theoretical stages to 8 the mole fraction of the methyl-acetate decreased to 0.051 with 10.2% conversion Figure 8. So when decreasing the number of stages to 4 stages the mole fraction decreased, also when increasing the number of stages to 8 stages the mole fraction decreased. Therefore, the best condition to produce methyl acetate and water for the given feed analysis is 6 theoretical stages.

Above the liquid holdup was assumed to be constant throughout the column. When the liquid holdup was changed methyl acetate composition in the distillate reached to a maximum of 0.252 at batch time between 0.4-0.8hr with 50.4% conversion Figure 9. While for constant liquid holdup the simulation curve reach to a maximum of 0.247 at batch time range 1.2-1.4hr. Therefore the effect of liquid holdup on the concentration of methyl acetate required less batch time than for constant liquid holdup.

Conclusions:

1- The results obtained from the developed model computer program give a good agreement with experimental and theoretical results obtained Eluge (16).

2- The liquid composition of the desired product methyl acetate at distillate section is 24.7%; with 49.4% conversion, while for water the mole fraction is 40%.

3- The effect of changing batch time on the product of methyl acetate is studied, while decreasing the batch time than 1.7hr, the liquid mole fraction of methyl acetate decreased to 22.5%, with 45% conversion. When increasing the batch time the liquid mole fraction of methyl acetate decreased to 21.7, with 43.4% conversion. The best condition for the design analysis of produced methylacetate was when batch time equals to 1.7hr.

4- The effect of changing number of stages product methyl acetate is studied, while decreasing the number of stages to 4 stages the liquid mole fraction of methyl acetate decreased to 5%, with 10% conversion, when increasing to 8 stages the liquid mole fraction of methyl acetate also decreased to 5.15%, with 10.2% conversion. The best condition for the design analysis of produced methylacetate was when number of stages equals to 6 theoretical stages.

5- Changing the liquid hold up throughout the tower will increases the overall conversion of methyl acetate to 50.4% and the recovery of methyl acetate increased to 92.5% and requires less batch time than for constant liquid holdup.

References

1- Sagar B., Malone M. F., and Doherty M. F., “Selectivity targets for batch reactive distillation" American Chemical Society, 2003.

2- Kian H. L.,” Optimal design and operation of batch distillation”, Computer Aided Process Engineering, 2003.

3- Sundmacher, and Steyer; ‘Batch Reactive Distillation for Recovery of Lactic Acid and Manufacture of Lactic Acid Esters”; 2002.

4- Mujtaba I. M. and Macchietto S., "Efficient Optimization of Batch Distillation with Chemical Reaction using Polynomial Curve Fitting Technique", Industrial and Engn. Chemistry Research, 36, 6, 2287-2295, 1997.
5- Aspen P.; "Unit Operation Models"; Version 10.2; 2000.
6- Huckaba and Danly, Am. Inst. Chem. Eng. J., 6,335; 1960 cited in [10].
7- Meadows, E. L., Chem. Eng. Symp. Ser. No. 46, 59, 48-55, 1963.
8- Distcfano G. P.," Mathematical Modeling and Numerical Integration of Multi component Batch Distillation Equations", AICE. Journal, January, 190 - 199, 1968.
9- Boston et al., "Foundations of Computer Aided Chemical Process Design, Vol. II, AIChE Journal, R.H.S. Mahand W.D Seider, Eds, 203-237, 1981.
10- Marta and Juan C., "Identification of Multi-component Batch Distillation Processes Using Subspace Methods" ; 2000.
11-Galindez H. and Fredenslund A.; "Simulation of Multi-Component Batch Distillation Processes", Comput. Chem. Engn. Vol. 12, No. 4, pp. 281-288, 1988.
12- Ping Z.; "Dynamic Simulation of Multiphase Batch Distillation"; 2002.
13- Jimenez L., Basualdo M.S. Gomez J.C. et al, " Non linear Dynamic Modeling of Multicomponent Batch Distillation: a case study " Braz. J. Chem. Eng. vol. 19, No. 3, 307-317, ISSN0104-632, July/ Sept. 2002.
14- Mujtaba I.M. and Macchietto S.,"efficient optimization of Batch Distillation with Chemical Reaction using Polynomial Curve Fitting Technique" Industrial and Eng. Chemistry Research, 36, 6, 2287-2295, 1997.
15- Lee J. W., Bruggemann S. and Westerberg A.W. "Visualization of Ethyl Acetate Reaction Distillation System" Conference proceeding, Technical Report March 2001.
16- Elgue S., Le Lann J.M., Cabassud M., Part L., Cezerac J., "Influence of the optimization approach on the determination of operating conditions of batch global syntheses", Chem. Biochem. Eng, 2003.
17- Hamodi H.M., Nakkash N.B. and Al-Habobi N. "Simulation of Batch Reactive
Distillation for Producing Methyl Acetate, MSc. Thesis, Al-Nahrain University,
Baghdad-Iraq, 2004.
18- Patel R., Singh K. Moses V. and Tade O. "Dynamic Simulation of Reactive Distillation Column for Ethyl Acetate Synthesis "Chemical Product and Process Modeling, Vol.2 Issue 2, Article 5, 2007.
19-Henely J.R and Seader J.D., "Equilibrium Stage Separation Operations in Chemical Engineering", John Wiley and Sons, 1981.

Nomenclatures

Btime	Batch time, hr
Cc_i	Concentration of each component i at any time
D	Distillate flow rate (Mole/hr)
Dens	Density (kg/m3)
E	Activation energy (kJ/kmol)
Gc	Constant volume holdup in the condenser, (m3)
Gp	Constant volume holdup in the Plates, (m3)
H	Enthalpy,(J/mol)
H^o_{fi}	Enthalpy of formation (J/mol)
HLi	Total enthalpy of liquid phase (J/mol)
H_{mix}	Enthalpy of mixing (J/mol)
H_v	Ideal gas state enthalpy (J/mol)
H_{va}	Ideal gas state enthalpy (J/mol)
L	Liquid flow rate, mole/hr
K	Distribution coefficient

K_{eq} Equilibrium constant for chemical reaction
K_{ester} Pre-exponential factor, 1/kgmol.min H2S04. l
Mc Liquid holdup in the condenser section, (mole)
Mn Liquid holdup on each tray, mole
Mt Liquid holdup in the reboiler section, (mole)
N Number of feed component.
Nt Total number of stages
P Pressure, mmHg
Q_r Inter-heater at the reboiler section, (kJ/min)
Rate Rate of reaction, mole/hr
Stime Step time, (sec)
T Temperature, (K)
V Vapor flow rate, (mole/hr)
vol Volume of one tray, (m3)
xi Liquid mole fraction.
X The experimental conversion (from reference (16))
yi Vapor mole fraction.

Greek Letter

γ Liquid phase activity coefficient.

Subscript

i Component i in the mixture
j Component i in the mixture at stage j

k Type of functional group
m Number of stages

Superscript

o Pure component
id Ideal solution
eq Equilibrium
mix Mixing

Table (1) Initial feed composition (16, 17)

Component	Mole Fraction
Methanol	0.5
Acetic Acid	0.46
Water	0.03
Methyl acetate	0.01

Table (2) Initial input data (16, 17)

Number of stages	6
Reflux Ratio	2.5
Liquid holdup reboiler, plates, and condenser	85, 0.5, and 14.5 moles.
Feed input	100mole
Batch Time	1.7hr
Operating pressure	1atm
Initial temperature	310.15K

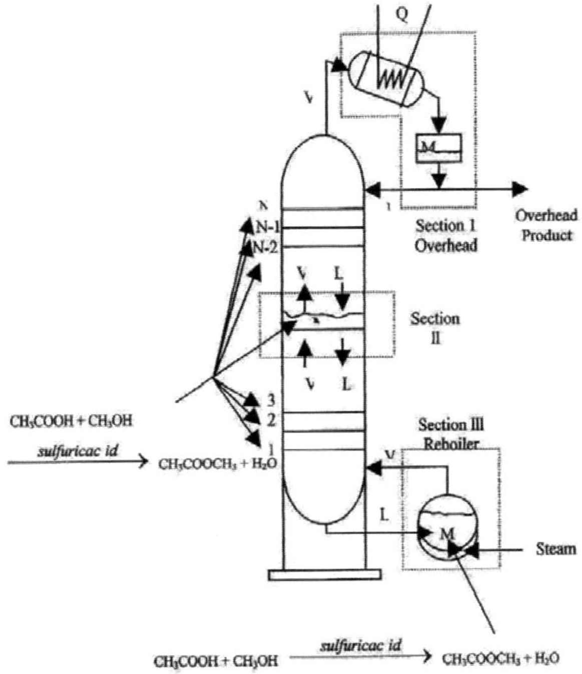

Figure 1: Batch Reactive Distillation

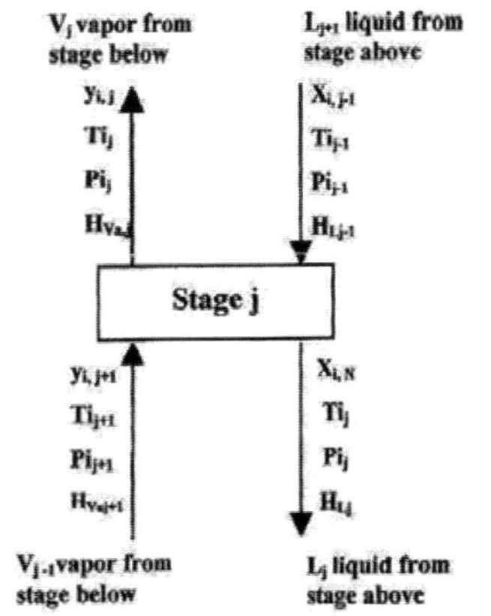

Figure 2: General Equilibrium Stage

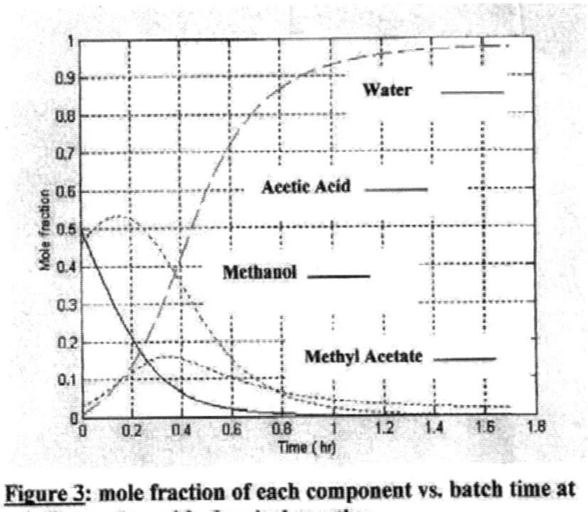

Figure 3: mole fraction of each component vs. batch time at reboiler section with chemical reaction

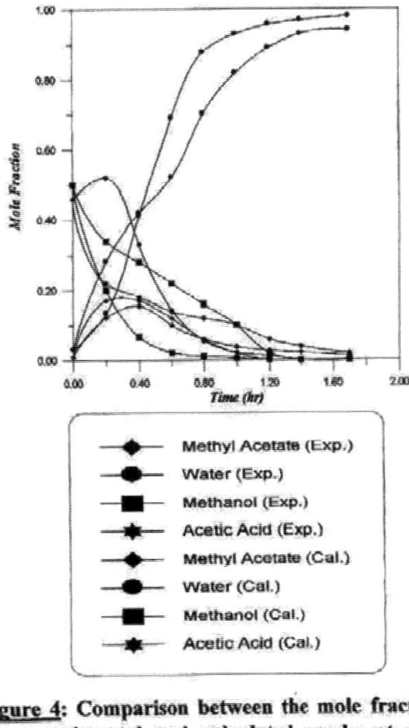

Figure 4: Comparison between the mole fractions of the experimental and calculated works at reboiler (with chemical reaction)

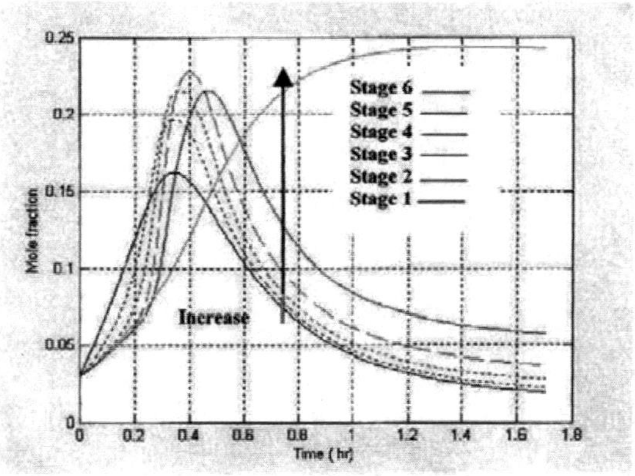

<u>Figure 5</u>: mole fraction of the desired product methyl acetate vs. batch time at each stage with chemical reaction

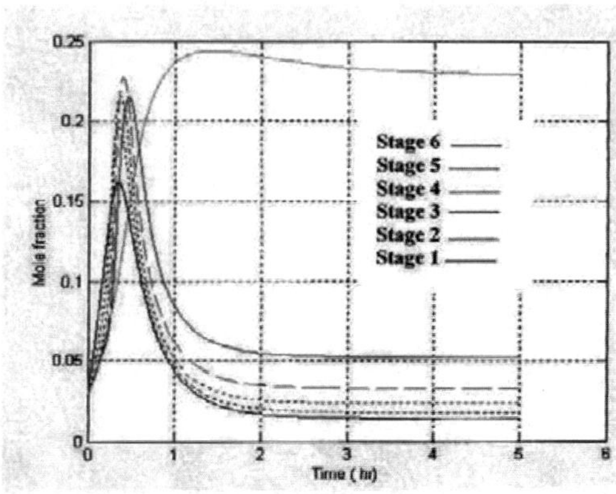

<u>Figure 6</u>: mole fraction of the methyl acetate vs. batch time at each stage (5hr)

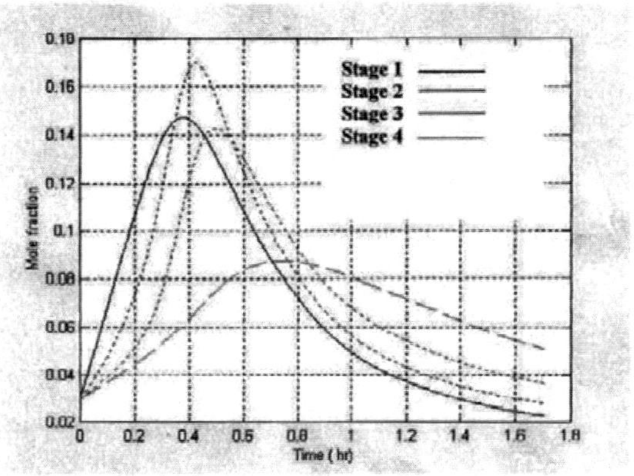

<u>Figure 7</u>: mole fraction of the product methyl acetate vs. batch time at each stage (4 stages)

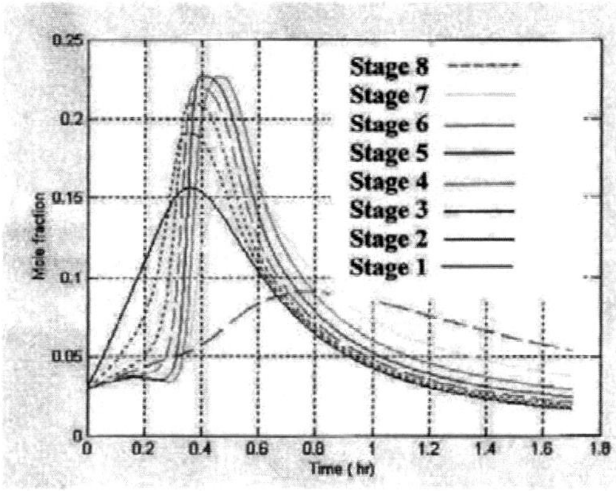

<u>Figure 8</u>: mole fraction of the product methyl acetate vs. batch time at each stage (8 stages)

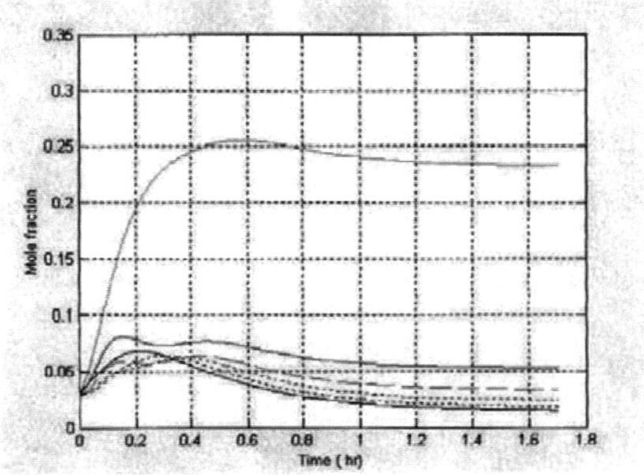

<u>Figure 9</u>: mole fraction of the product methyl acetate vs. batch time at each stage

(31g) A New Separation Technique – Distillative Crystallization

Lie-Ding Shiau, Department of Chemical and Materials Engineering, Chang Gung University, Taoyuan, Taiwan

Keng-Fu Liu, Department of Chemical and Materials Engineering, Chang Gung University, Taoyuan, Taiwan

Distillative crystallization (DC) is a new separation technique which combines distillation and crystallization to separate the mixtures with the close boiling temperatures. Basically, DC is operated at a triple-point condition, in which the liquid is simultaneously vaporized and crystallized due to the three-phase equilibrium. In essence, the process is continued until the liquid phase is completely eliminated and only the pure solid crystals remain in the feed. Thus, crystal washing is not required since only pure solid crystals remain in the feed and no impurities are adhered on the crystal surfaces at the end of DC. In the previous research, DC has been successfully applied to separate p-xylene from the mixed xylenes by Shiau et al. [1-3].

DC is introduced in this work to purify styrene (ST) from the mixture consisting of styrene (ST), ethylbenzene (EB), p-xylene (PX), m-xylene (MX) and o-xylene (OX). The experiments show that the purity of ST can be purified from 0.85 to 0.98 or from 0.90 to 0.99 in the mixture consisting of ST, EB, PX, MX and OX. Thus, DC provides an efficient separation method to separate ST from the mixture consisting of ST and the mixed xylenes. The recovery rate of ST in the experiments can be greatly improved by adjusting the operating temperature and pressure during the DC process.

[1] Shiau, L. D.; Wen, C. C.; Lin, B. S., 2005, "Separation and purification of p-xylene from the mixture of m-xylene and p-xylene by distillative freezing", Ind. Eng. Chem. Res. , 44(7), p.2258-2265

[2] Shiau, L. D.; Wen, C. C.; Lin, B. S., 2006, "Application of distillative freezing in the separation of o-xylene and p-xylene", A.I.Ch.E. Journal, 52(5), p.1962-1967.

[3] Shiau, L. D.; Wen, C. C.; Lin, B. S., 2008, "Separation of p-xylene from the multicomponent xylene system by stripping crystallization", A.I.Ch.E. Journal, 54(1), p.337-342.

See more of this Session: Advances in Distillation & Absorption I
See more of this Group/Topical: Topical 8: Distillation Topical

(55a) Novel Graphical Design Methods for Complex Membrane Configurations and Hybrid Processes

Mark Peters, Wits University, Johannesburg, South Africa

Diane Hildebrandt, Comps, Wits University, Johannesburg, South Africa

David Glasser, Comps, Wits University, Johannesburg, South Africa

Shehzaad Kauchali, Wits, Wits University, Johannesburg, South Africa

Shortcut methods for the design and synthesis of separation systems are useful, especially in the conceptual stages of a process design. However, current approaches incorporated in these techniques result in them being suitable for traditional designs only, and unable to manage novel or complex configurations. Using column profile trajectories to describe the compositional behaviour of material in both distillation and membrane sections, it will be shown how any configuration, no matter how complex, can be modelled. With this comes an overall deeper understanding of the operation of the chosen system. As an example, a Petlyuk-type arrangement will be considered – firstly incorporating membrane permeation units as the only means of separation, and then secondly linking membrane units with distillation thereby forming a hybrid. The feasibility of each design is discussed, and compositional regions of feasibility are established, where necessary. Although the nature of column profiles may appear to be complicated, this method allows one to quickly and easily understand complex configurations, thus saving time, money and potentially reducing energy consumption.

See more of this Session: Advances in Distillation & Absorption II

See more of this Group/Topical: Topical 8: Distillation Topical

(55b) HIDiC as A Retrofit Technology

Zarko Olujic, Process & Energy, Delft University of Technology, Delft, Netherlands

S. Tromp, Process & Energy, Delft University of Technology, Delft, Netherlands

This paper explores potential for implementation of an internally heat integrated distillation column (HIDiC) as a retrofit option, taking an actual heat pump assisted PP-splitter as the base case. The retrofit option is compared with a new VRC designed to achieve the same capacity increase by placing new unit in parallel to existing VRC. Present feasibility study indicates that a capacity increase of 70% can be realized with energy savings up to 25%, which however can not be fully realized due to heat transfer area limitation. In a feasible design the potential energy saving gain is halved. Nevertheless, regarding the scale of application, the estimated capacity increase and energy saving percentages translate into impressive financial gains, and this will hopefully help to increase chances for HIDiC to be considered by chemical process industries as a prospective debottlenecking and energy-saving technology.

See more of this Session: Advances in Distillation & Absorption II

See more of this Group/Topical: Topical 8: Distillation Topical

(55c) Reboiler Circuits for Trayed Towers - Part 1

Ed Grave, ExxonMobil Upstream Research Company, Houston, TX

Todd Marut, ExxonMobil Research and Engineering Company, Fairfax, VA

Daniel R. Summers, Sulzer Chemtech USA, Inc., Tulsa, OK

Jeffery A. Bell, UOP LLC, Des Plains, IL

Waldo E. De Villiers, Shell Global Solutions (US) Inc., Houston, TX

Henry Z. Kister, Fluor Corporation, Aliso Viejo, CA

Paul Morehead, Koch-Glitsch, LP, Dallas, TX

Ron Olsson, Celanese Ltd., Bishop, TX

Joseph Flowers, E.I. DuPont de Nemours & Company, Wilmington, DE

Joseph Parker, Eastman Chemical Company, Kingsport, TN

Simon Xu, Process Technology, Shaw E&C, Houston, TX

Larry Wilder, Dow Chemical Company, South Charleston, WV

Doug Bouck, Separations Solutions, Ltd., Solon, OH

John P. Farone, Dow Chemical (retired), Cedar Park, TX

Neil Yeoman, LNCK Associates, Merrick, NY

The reboiler generally supplies most of the energy required to effect component separation. If too much heat is supplied, the tower will flood; conversely, if too little heat is available, separation performance decreases via poor reflux ratio (pinching), excessive weeping or poor tray action. Proper design of reboiling systems involves coordinating aspects both outside and inside the tower. This Design Practice focuses on both of these aspects to ensure proper operation of the overall reboiler system. Important literature discussing these topics is also cited.

See more of this Session: Advances in Distillation & Absorption II
See more of this Group/Topical: Topical 8: Distillation Topical

(55d) Reboiler Circuits for Trayed Towers - Part 2

Ed Grave, ExxonMobil Upstream Research Company, Houston, TX

Todd Marut, ExxonMobil Research and Engineering Company, Fairfax, VA

Daniel R. Summers, Sulzer Chemtech USA, Inc., Tulsa, OK

Jeffery A. Bell, UOP LLC, Des Plains, IL

Waldo E. De Villiers, Shell Global Solutions (US) Inc., Houston, TX

Henry Z. Kister, Fluor Corporation, Aliso Viejo, CA

Paul Morehead, Koch-Glitsch, LP, Dallas, TX

Ron Olsson, Celanese Ltd., Bishop, TX

Joseph Flowers, E.I. DuPont de Nemours & Company, Wilmington, DE

Joseph Parker, Eastman Chemical Company, Kingsport, TN

Simon Xu, Process Technology, Shaw E&C, Houston, TX

Larry Wilder, Dow Chemical Company, South Charleston, WV

Doug Bouck, Separations Solutions, Ltd., Solon, OH

John P. Farone, Dow Chemical (retired), Cedar Park, TX

Neil Yeoman, LNCK Associates, Merrick, NY

The reboiler generally supplies most of the energy required to effect component separation. If too much heat is supplied, the tower will flood; conversely, if too little heat is available, separation performance decreases via poor reflux ratio (pinching), excessive weeping or poor tray action. Proper design of reboiling systems involves coordinating aspects both outside and inside the tower. This Design Practice focuses on both of these aspects to ensure proper operation of the overall reboiler system. Important literature discussing these topics is also cited.

See more of this Session: Advances in Distillation & Absorption II
See more of this Group/Topical: Topical 8: Distillation Topical

(55e) Gamma Scanning of FRI Reboiler Vapor Return Lines

Lowell Pless, Tracerco, Pasadena, TX

Ron Carlson, Tracerco, Pasadena, TX

Mike Resetarits, Fractionation Research, Inc., Stillwater, OK

A. Y. Ogundeji, Fractionation Research, Inc., Stillwater, OK

FRI's distillation unit includes two columns and two reboilers. Those reboilers are virtually identical and can be linked together. Liquid entrainment issuing from the tops of those reboilers is difficult to calculate due to the presence of simple knock-down baffles just beneath the vapor product nozzles. A window directly opposite the low-pressure column reboiler vapor return nozzle is supposed to allow easy visual observation of the reboiler entrainment. During November 2009 testing, a rain of condensate inside that column made it very difficult to see the entrainment level. Gamma scanning, in the form of monitoring with nucleonic density gauges, of the two reboiler vapor lines revealed that entrainment from the two reboilers was lower than expected, even at very high heat duties.

KEYWORDS

distillation reboiler scanning entrainment

See more of this Session: Advances in Distillation & Absorption II
See more of this Group/Topical: Topical 8: Distillation Topical

Diagnostics of High Pressure Depropanizer

Lowell Pless

Tracerco
4106 New West Drive
Pasadena, TX 77507, USA
Phone: 1-281-291-7769
Email: Lowell.Pless@tracerco.com

Christopher McKeown and Simon Xu

Shaw Energy & Chemicals
1430 Enclave Parkway
Houston, TX 77077, USA
Phone: 281 368 4176
Email: Christopher.Mckeown@shawgrp.com
Simon.Xu@shawgrp.com

Presented in Distillation Topical Session of the 2010 AIChE Spring Meeting, San Antonio, TX, March 21-25, 2010

UNPUBLISHED

Diagnostics of High Pressure Depropanizer

Lowell Pless
Tracerco

Christopher McKeown and Simon Xu
Shaw Energy & Chemicals

Distillation Topical Session of the 2010 AIChE Spring Meeting, San Antonio, TX, March 21-25, 2010

Abstract

A Depropanizer tower with a large diameter top section and smaller diameter bottom section was field tested and studied for its root cause of flooding. In the diameter transition zone, a two-phase feed was being introduced. Gamma scans identified that the trays in the top section were flooding and liquid accumulation initiated in the transition zone. A neutron backscatter technique was used to investigate downcomer hydraulics and identify the root-cause of flooding. Coupled with the field troubleshooting, a careful analysis of the design revealed likely causes of the flooding problem. Recommendations for modifications were made and installed.

The High Pressure (HP) Depropanizer was a replacement tower as part of a 30% expansion of an ethylene plant. On start-up, plant capacity was limited to 96% of design due to the performance of the HP Depropanizer – C4's in the overhead would go out of specification. As well when the plant approached 96% capacity sharp increases in HP Depropanizer ΔP was observed.

The HP Depropanizer is not a typical distillation tower design as it is part of a heat pump system. As such it does not have the typical overhead reflux arrangement. Instead of the overhead passing through a condenser and part of the liquid used as tower reflux, the HP Depropanizer overhead passes through a feed/effluent heat exchanger, a compressor, reactor train, chillers, condensers and finally to a reflux drum from where some liquid is used as tower reflux and the rest is fed to downstream process. See Figure 1.

The feeds to the HP Depropanizer represent the entire unit ethylene production so the tower being a bottleneck was an extremely troubling problem.

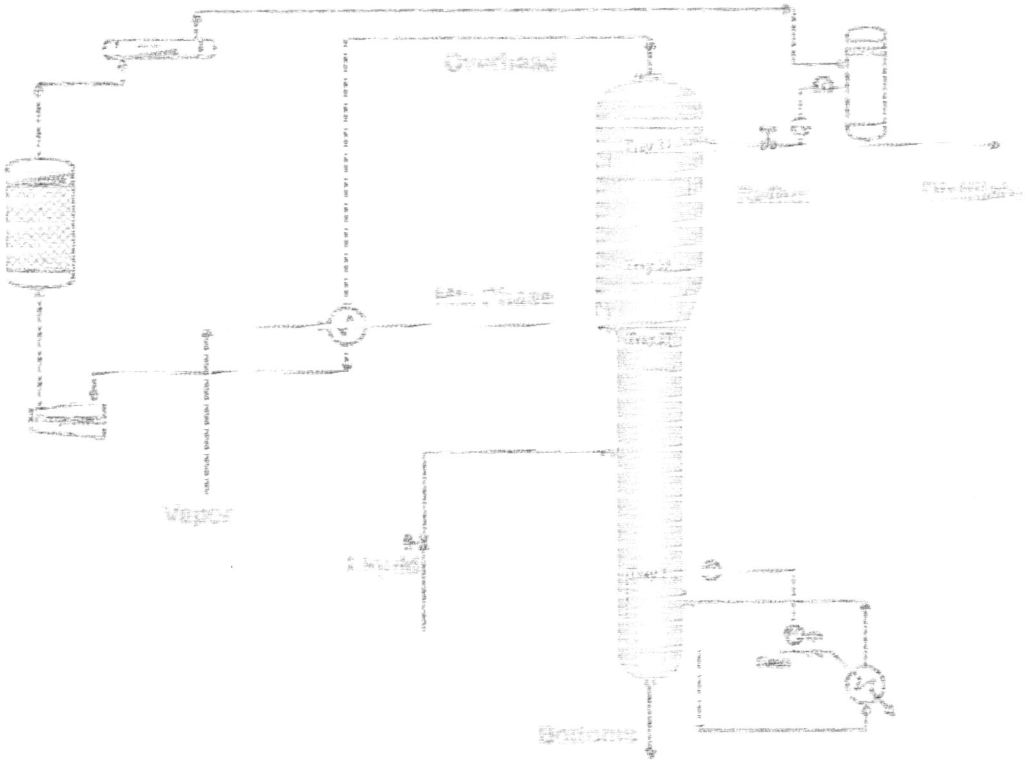

Figure 1. Process Scheme of HP Depropanizer

Initial Evaluation

The following shows process data concerning the operating problems with the HP Depropanizer:

When the feed load was at 96% of design rate,

ΔP across demister	0.016 kg/cm^2 (0.23 psi)
ΔP across top trays	0.07 kg/cm^2 (1 psi)
Butadiene in overhead	100 to 200 ppm mol

When the feed was increased above 96%,

ΔP across demister	0.03 kg/cm^2 (0.43 psi)
ΔP across top trays	0.13 kg/cm^2 (1.8 psi)
Butadiene in overhead	700+ ppm mol

Based on the process data above it seemed obvious that the top section of trays in the HP Depropanizer was flooding. Rigorous checks of plant operating data versus simulation calculations showed no major deviations from the process design. A gamma scan was performed to confirm the flooding, as well as to see where the flooding was originating.

Gamma scanning is a tower diagnostic technology using a small, gamma ray emitting radioactive source and a very sensitive radiation detector. The fundamentals of gamma scanning are based on the following relationship, Beer's Law, between material and radiation:

$$I = I_o\, e^{-\mu \rho x}$$

where I_o is the initial gamma ray intensity measured at a given distance with no material interfering with the radiation transmission, I is a radiation intensity measurement taken through a material, μ is the absorption coefficient specific to the material being measured, ρ is the material density, and χ is the material thickness. Thus, radiation intensity passing through a material decreases exponentially as a function of the material density. It should be understood that there are many conditions that need to be fulfilled in order for Beer's law to be valid, so typically Beer's law does not exactly hold valid for the calculation of density from typical gamma scan data. Nevertheless count rate has proven to be a useful parameter to approximately correlate to the average material density (ρ) between the gamma source and the detector. The correlation can be qualitatively expressed as,

$$\rho \propto \frac{1}{\ln\,(count\ rate)}$$

The formula indicates that the higher the count rate, the lower the average process density (ρ), where count rate is the gamma signal rate counted by the detection system.

When applied to distillation column scanning, a radiation source and detector are aligned on opposite sides of the column. The scanning apparatus is kept on the outside surface of the process equipment, so the procedure is non-invasive and non-interactive to the normal operation of the process. From the top of the column the source and detector are lowered synchronously while radiation intensity measurements are recorded at specified intervals. The "material" being measured is the composite thickness and density of the column walls, column insulation, column internals, and the process material inside the column. For a given column the physical "material" – walls, insulation, etc. have a fixed thickness. In situations where the physical "material" thickness has known changes, such as column diameter and wall thickness changes, the radiation intensity measurements can be adjusted for these factors. Otherwise detectable differences in radiation intensity will be a function of the density changes of material inside the piece of equipment being scanned. When the radiation intensity detected is low, the radiation is passing through higher density material (metal or liquid holdup); and when the radiation intensity is high, the radiation is passing through lower density material (vapor spaces). Based on the characteristics of the density profile and what is known about the column internals and service, diagnostic conclusions can be made regarding the column's hydraulic performance and the physical condition of its internals.

The gamma scan confirmed that the top section of the HP Depropanizer was flooding. As seen in Figure 2 the flooding started at Tray 28 indicating that the downflow liquid from Tray 28 was restricted at the transition zone. However the initial scan could not provide sufficient information to answer the questions on the root cause of flooding:

- Was the flooding from entrainment or downcomer backup?
- What was restricting the flows?
- How to solve the flooding problem?

More field tests and process and hydraulic analysis were required to understand the flooding mechanism and the root causes.

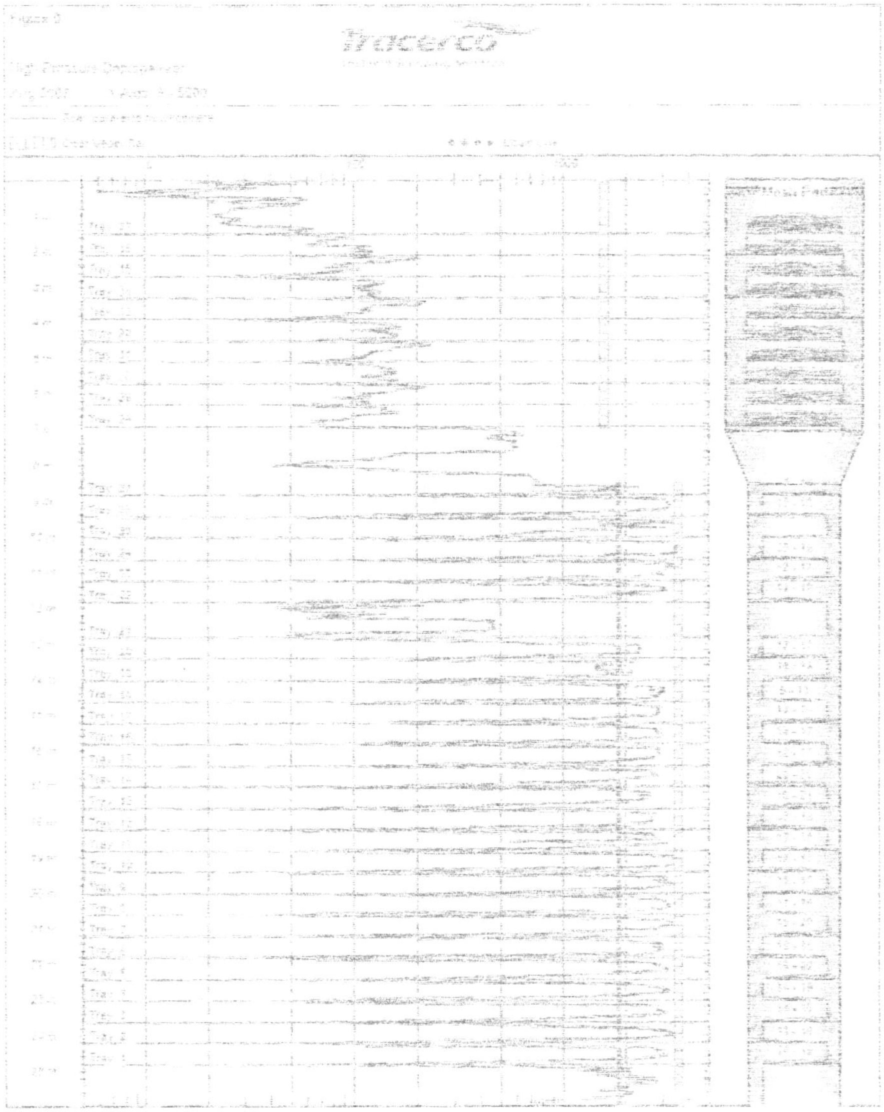

Figure 2. Initial Scan Showed that Flooding Started at Tray 28.

More Field Tests – Gamma Scans

To investigate the hydraulic abnormalities in the transition zone, two more gamma scans were performed, one for active areas and one for downcomers, as shown in Figure 3. Scan 1 was performed across the active area by scanning parallel to the downcomer; Scan 2 was perpendicular through the downcomer.

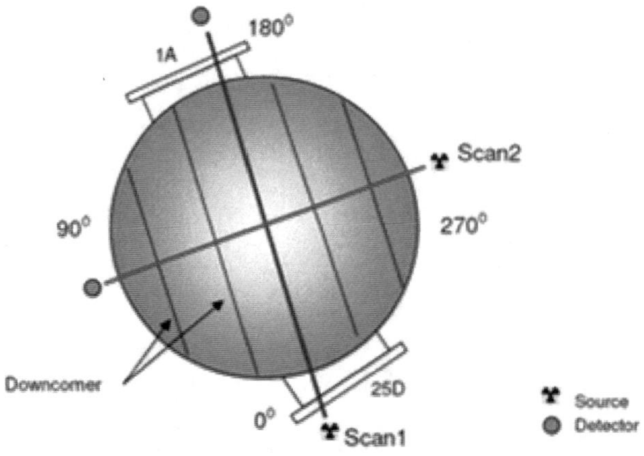

Figure 3. Scan Line Orientations

Figure 4 is an expanded view of the two scans through the transition section. While the orientation of Scan 2 was not perfect for investigating the hydraulic condition of Tray 28's downcomer it did reveal some new information.

Scan 1 showed a level of aerated liquid on Tray 27 while Scan 2 showed Tray 27 operating dry, as seen in Figure 4. Within the highlighted area on Figure 4 Scan 1 (the blue curve) showed a response at Tray 27 signifying some dense material (liquid) on the active area of Tray 27. Conversely Scan 2 (red curve) within the highlighted area showed nearly no response at Tray 27. There was no dense material i.e., no liquid holding on Tray 27. Since Tray 27 had been seen holding liquid there was no question that Tray 27 was intact mechanically and capable of holding liquid. Instead during the operation when Scan 2 was performed, while the flooding persisted Tray 27 had gone dry, presumably because liquid downflow from Tray 28 above was severely restricted.

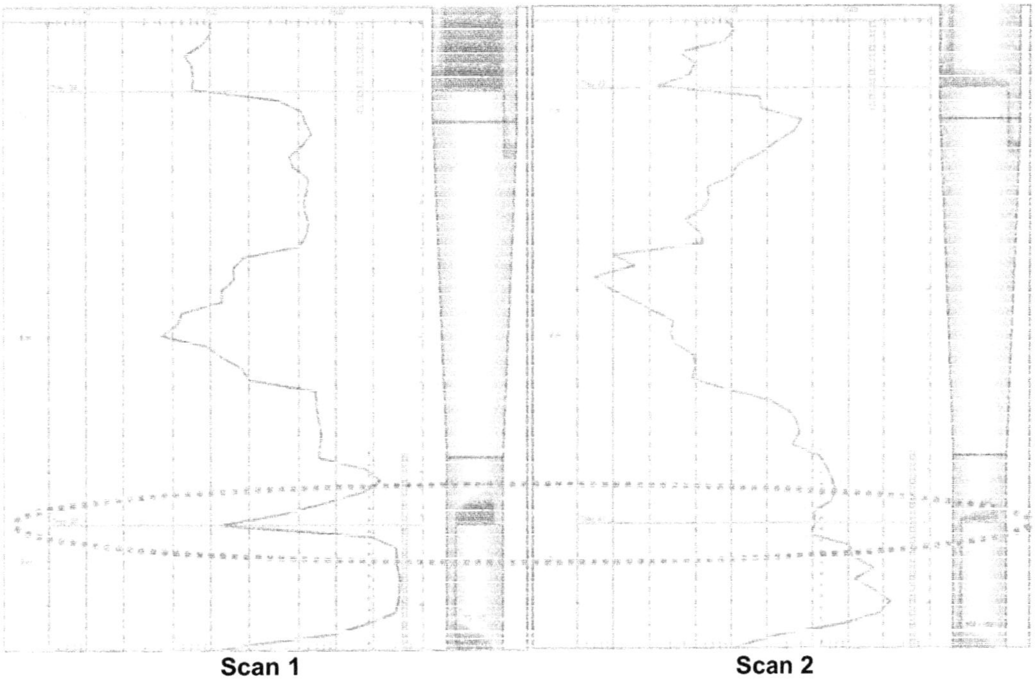

Figure 4. The Liquid Holdup or Froth on Tray 27 Appeared Unsteady During the Scans

Reconciling these results with plant process data (Figure 5) showed that as the pressure drop or flooding severity increased and prolonged, Tray 27 dried up, again confirming that the downcomer from Tray 28 was restricted.

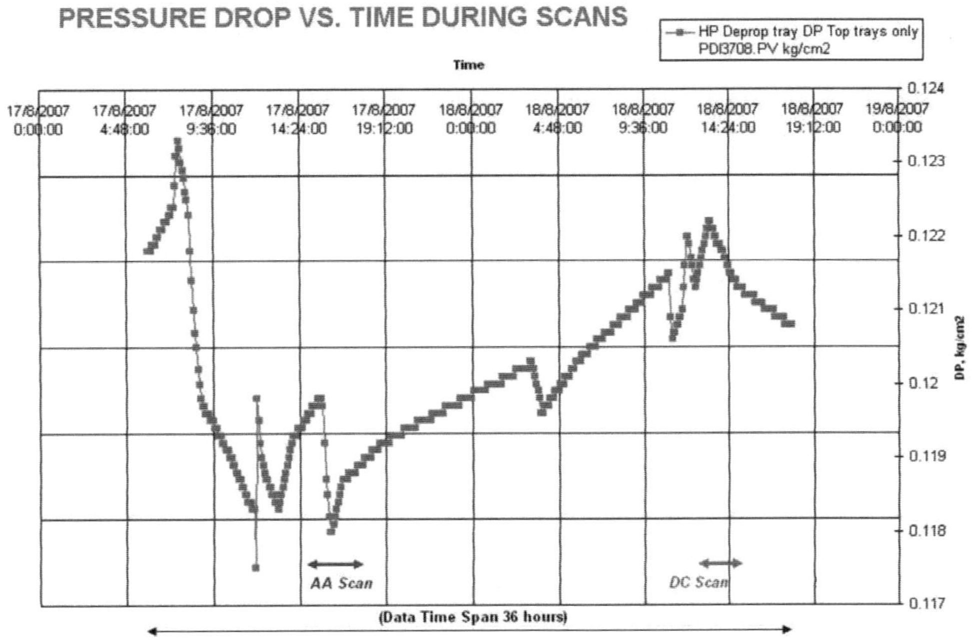

Figure 5. Pressure Drops (Indication of Flooding Severity) during the Scans

Review of Tower Internals

Several aspects of the design of the HP Depropanizer were studied, focusing on the feed transition area where the flooding initiated. Three items were identified that, listed by relative importance, could be causing the flooding of the HP Depropanizer. Please refer to Figure 6.

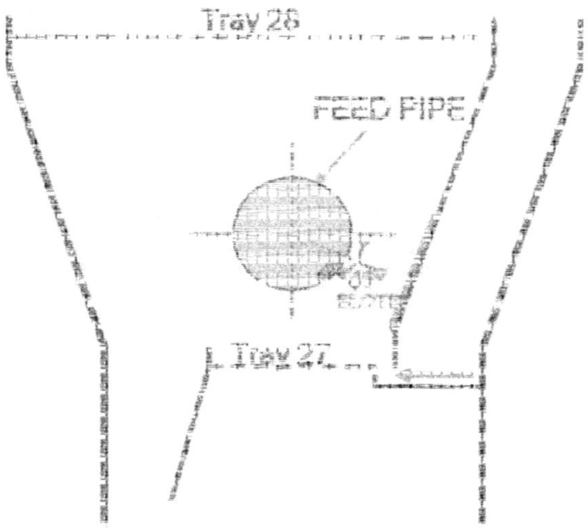

Figure 6. Tower Internals of the Transition Section

The process simulation showed a mixed phase feed of approximately 93 wt% vapor and 7 wt% liquid. The 2-phase pipe distributor was designed for the exit slots to be centered at an angle of 60° from vertical. Unfortunately some construction drawings showed that the pipe distributor was installed with the exit slots centered 45° from vertical. Therefore, incoming feed was being directed towards the bottom of Tray 28 downcomer.

The downcomer sump of Tray 28 had been sized based on the liquid load in the middle section of the tower rather than the liquid load coming from the top section. For this reason the downcomer through the transition area was oversized. Because of this the liquid head in the downcomer was lower than expected and gave some concern that the liquid seal could be lost allowing vapor feed to travel up the downcomer. A second goal from the scanning activity was to try to determine if this vapor-bypassing was happening.

An impingement plate is typically used on the downcomer apron to act as a heat insulator to avoid downcomer liquid from being vaporized by the warm feed. In this design the feed temperature was only 3°C (5°F) warmer than the downcomer liquid so the impingement baffle had been left off. In reality the actual temperature profile may not be exactly as predicted by simulation models, as process turbulences could cause temperature deviations. An impingement plate should be installed as an extra precaution.

Evidence was accumulating that vapor flow from the feed was going up the downcomer from Tray 28. At this point everyone involved, especially plant management, wanted "proof-positive" that the hypothesis of vapor-bypassing up Tray 28's downcomer was actually happening.

Verifying the Vapor-Bypassing by Neutron Scans

Under most circumstances gamma scanning through tray downcomers is the recommended approach. However there were some extraordinary reasons why this was not the best approach in the circumstance of this HP Depropanizer.

The physical structure of the HP Depropanizer, top large diameter and bottom smaller diameter, is challenging for gamma scanning, as shown in Figure 7. As the gamma source and detector travel down through the transition zone, the source and detector are essentially hanging "in the air", instead of maintaining contact against the walls. Due to limited access through the transition zone this was unavoidable on this tower.

Additionally the diameter is constantly changing through the cone-shape transition zone. Even if one was able to keep the source and detector up against the column wall, for each drop in elevation the radiation counts at the detector will vary based on the constantly changing diameter. It could be possible, even if improbable, that a change in internal process density would be negated or exaggerated by the change in diameter or distance from source to detector.

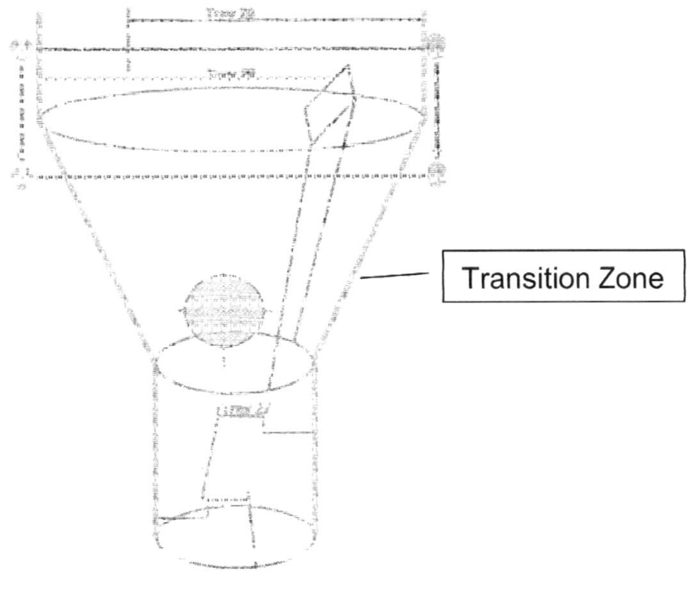

Figure 7. The Cone-Shape Transition Section Imposed Challenges to Gamma Scans

Ideally the best way to gamma scan tray downcomers is to scan parallel to the downcomer walls while the radiation beam passes through the middle of the downcomer. Given the physical features discussed above coupled with the relatively narrow downcomer area, a gamma scan was not going to provide the kind of detail information desired about the operating condition of the downcomer through the transition zone.

The best approach to scan through the downcomer in the transition zone would be to scan with neutrons. Neutrons essentially measure the hydrogen concentration of the material in front of the neutron source/detector apparatus. In this case we would expect a large response where hydrocarbon liquid was present and a smaller response where vapor or less liquid was present.

> The neutron scan device consists of a fast neutron source and a slow neutron backscatter detector. Fast neutrons from the source collide with hydrogen nuclei in the material inside the vessel being scanned. The collisions transfer some of the energy to the hydrogen nuclei, reducing the neutrons' speed and changing their direction of travel in a process known as neutron moderation. Some of the affected neutrons are deflected back (backscatter) toward the detector where they are received and counted. A high concentration of hydrogen nuclei in the material adjacent to the source will result in a large number of slow neutrons detected. In other words, the neutron backscatter technique is basically detecting the "concentration" of hydrogen atoms in the process material adjacent to vessel walls. Compared with gamma scans, the neutron backscatter device is easy to use and can usually be operated by a single person.
>
> The neutron backscatter technique can be used to locate interfaces between materials having different concentrations of hydrogen atoms. In a storage tank, it can differentiate between the interfaces of vapor, oil, water, and sludge. In a column having trays with downcomers, it can determine the froth height or liquid backup in downcomers. It is particularly useful for measuring the liquid height on chimney trays, in draw sumps, in downcomers, or in tower bases without having to do a full column scan.

The neutron scan started above the opening of Tray 28's downcomer and proceeded down to below the bottom of this downcomer. The black curve in Figure 8 shows the results from this scan. At the bottom sump of the downcomer where one would expect liquid the neutron response instead showed vapor. There was a "liquid" response further up in the downcomer, several centimeters from the bottom. The belief is this was a layer of highly aerated liquid or foam suspended in the downcomer by the vapor. Furthermore at the same time neutron readings were taken on the active area of Tray 27 and it was seen to be void of any substantial level of liquid.

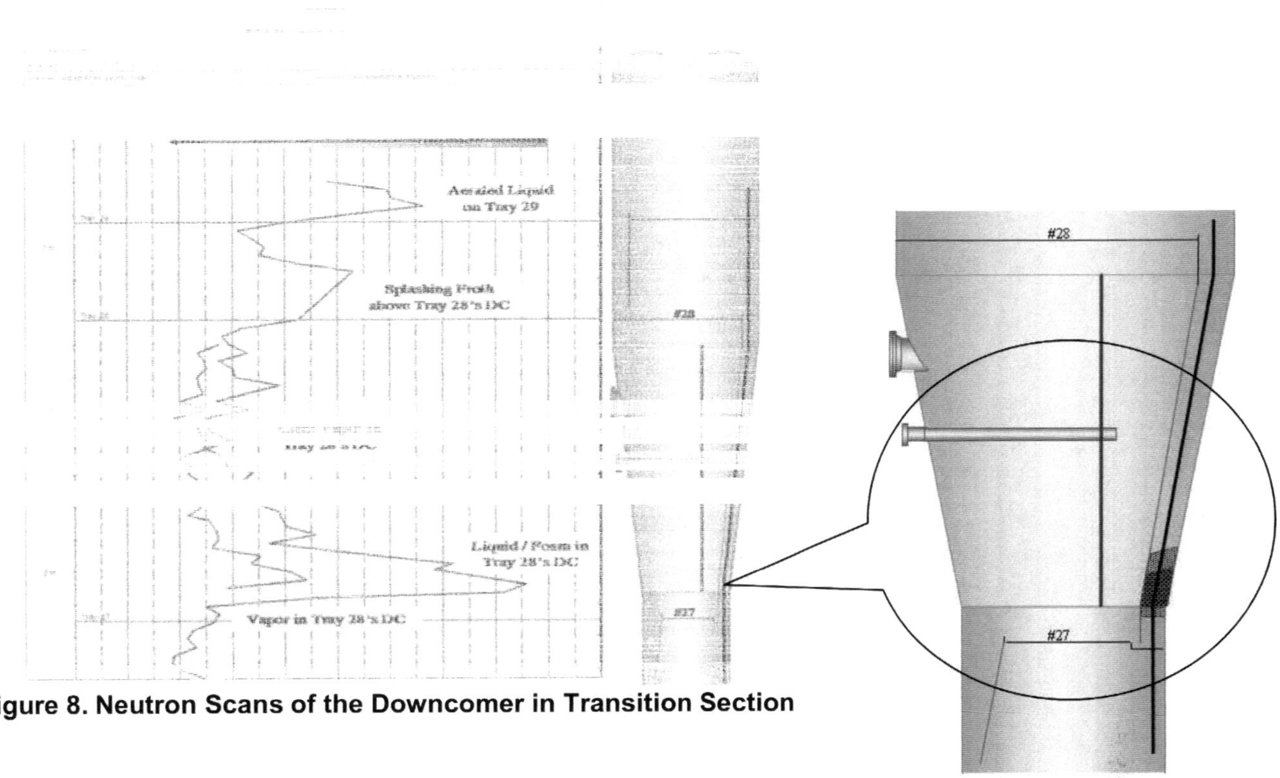

Figure 8. Neutron Scans of the Downcomer in Transition Section

Thus the neutron scans validated the hypothesis that vapor was blowing back up Tray 28's downcomer and preventing liquid from down-flowing out of the top section, thus flooding the top of the HP Depropanizer.

Solve the Flooding Problem

With a clear picture on the root cause of flooding in the transition section, several steps were proposed and installed to correct the configuration of the HP Depropanizer feed zone (refer to Figure 9):

- First, the feed pipe slots were modified to be 60° from vertical, reducing the probability of vapor blowing directly onto the downcomer outlet,
- Second, the slot open area was increased to reduce the vapor exit velocity,
- Third, an impingement plate was added to the downcomer apron to provide insulation between the hot vapor and cooler liquid to avoid downcomer liquid vaporizing,
- Fourth, the impingement plate had a ledge added to its bottom edge to further guard against the vapor impinging down onto the liquid leaving the downcomer,

- Fifth, the under downcomer clearance was reduced in order to increase the liquid head in the downcomer to secure a liquid seal at the bottom of the downcomer.

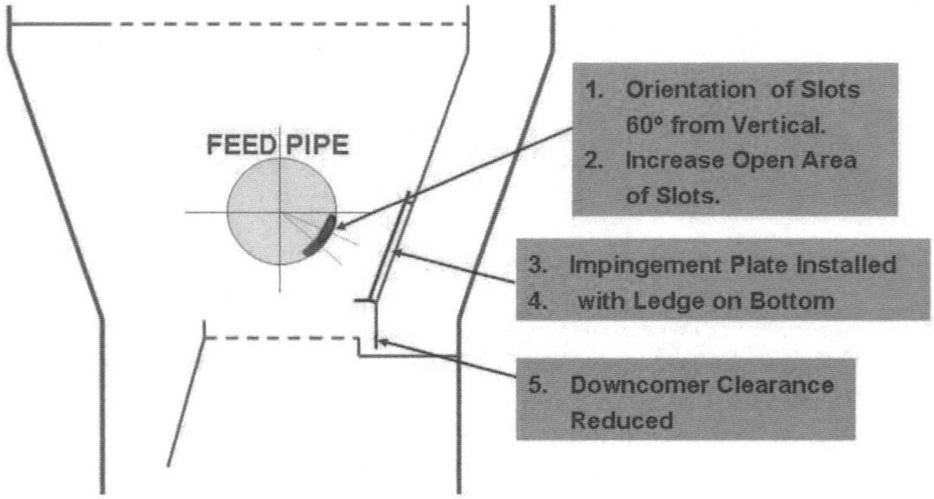

Figure 9. Modifications of the HP Depropanizer Transition Zone

It could be argued that not all of the above design changes were necessary to correct this flooding issue. However, all the changes recommended were low-cost and due to the importance of this tower every opportunity to improve the design was taken to ensure success following the plant shutdown.

After these modifications were installed the HP Depropanizer and the unit was restarted. The unit was able to demonstrate design capacity in a successful plant test run with no further symptoms of flooding.

Use of Gamma Scan in Obtaining Hydraulic Data for Trayed and Packed Columns

A. Shariat*, T. J. Cai

Fractionation Research, Inc., 424 S. Squires St. Suite 200, Stillwater, OK 74076, USA

To be Presented in the Advances in Distillation & Absorption II of the 2010 AIChE Spring Meeting

San Antonio, Texas

March 21-25, 2010

UNPUBLISHED

Use of Gamma Scan in Obtaining Hydraulic Data for Trayed and Packed Columns

A. Shariat*, T.J. Cai

Fractionation Research, Inc.

424 S. Squires St, Suite 200

Stillwater, OK 74074 USA

ABSTRACT

Gamma ray absorption has been used for troubleshooting operating distillation and absorption columns. In such an application the scan would point to the missing or collapsed trays, or lack of vapor in a space between two trays indicating the flooding condition. FRI does not use gamma scanning for such purpose, but it uses a high strength gamma ray source to determine quantitatively the froth density and liquid holdup in downcomer and on trays as well as the liquid holdups in random and structured packing columns. This paper shows experimental results for froth density profile in a sieve tray downcomer with iso-Butane/ n-Butane system at 11.4, 20.7, and 34.5 bar; and Cyclohexane/Normal heptane at 1.62 bar. The profiles of a straight downcomer and sloped downcomer are also compared.

Key Words: Distillation, sieve tray, gamma scan, froth density profile, liquid holdup, sloped downcomer, straight downcomer.

---

*Corresponding author. E-mail: shariat@fri.org , Phone: 1-405-385-0354, Fax: 1-405-385-0357

INTRODUCTION

To analyze the hydraulic behavior of trayed and packing columns, one needs to be able to measure froth density and holdup directly. Froth density is a key parameter of downcomer backup and the tray pressure drop. Holdup measurement in the packed columns was limited to the small diameter columns. Previously these parameters were measured directly using manometers and visual observations. The ability of monitoring and measuring these parameters on continuous basis, enhances the analysis and understanding of hydraulic performance of both trayed and packed columns.

Radioisotopes have been used since 1960s in detecting the hydraulic behavior during column operation. Most applications have been qualitative in nature and a diagnostic tool to determine the cause of column failures. The modern radioisotope survey methods are either non-intrusive, which use a sealed container as the radiation source; or intrusive where an unsealed radiation source is used. The sealed source methods include gamma scans, CAT Scans (Computer Aided Tomography), and Neutron Backscatter. The unsealed source methods, commonly called radioactive tracer tests, are either pulse or continuous injection. These methods have their own specific application, non-intrusive methods are the preferred method because they do not interfere with the column operation and do not affect the product quality.

The first application of gamma scan in hydraulic studies was reported by Beckman *et al* [1] who concluded the liquid flow is not absolutely uniform in a packed bed. The experimental setup was consisted of a 305 mm (12 in) glass column with ceramic packing with air water system. The gamma radiation source was very low strength mercury 203 with 13 millicuries strength. Such low strength cannot be applied to commercial size columns due to high absorptivity of the steel walls, and the fact that the radiation spreads causing the weakening of the radiation strength with the square of distance from the source.

On the commercial scale, gamma scanning has been used since 1960s[2] in column trouble shooting. In these applications, not only shutdown is avoided, but the ability to diagnose the column during operation, enables one to correctly identify the problem source. Trying to identify the column failures by shutting down and opening the column, may not render the actual conditions under which the column had failed. One of the drawbacks in applying the commercial application is use of low strength radiation source, which is a requirement for the safe transportation of the radioactive materials. The low strength radiation source results in a more qualitative scans and may not correctly identify the column problems; in such case, other radioisotope methods may be more appropriate for trouble shooting.

The radiation source used in the commercial gamma scanning should have sufficient photon energy to penetrate the column walls, the process fluid, hardware in the scanline and reach the detector in the opposite side of the column. The strength is measured in terms of KEV, Kilo Electron Volts. The most common sources that are in general use are Cesium-137 (662 KEV, half life of 30.07 years), Cobalt-60 (1170 and 1330 KEV, half life of 5.27 years). Half life is the time required for the strength to fall to half of its initial strength, due to decay. Since Cobalt-60 has higher energy, it can be used to scan thicker and larger diameter columns. Sodium -24 (1370 and 2750 KEV, half life of 15 hours) has been used for extremely large diameter column and/or very thick column walls.

As gamma photons travel through the tower and process fluid, they lose their energy by either Photoelectric Effect or Compton Effect. In the Photo Electric mechanism, the gamma ray photon strikes an orbital electron of the atom of the material that it is passing through, passes its total energy to the colliding electron and ceases to exist. In Compton Effect the penetrating electron collides with a valance electron, it only loses part of its initial energy and is deflected in direction other than when it approached the atom. This process may be repeated several times, and each time the penetrating photon passes part of the remaining energy to the colliding electron and changes direction, thus the term Compton Scatter. For a high resolution scan the weakened gamma ray should be rejected because they have nothing to do with the actual density of the material. With low strength source, due to environmental safety concerns and the existing technology, filtering the Compton Effect gamma rays is not possible. This introduces uncertainty in the scan that must be taken into consideration in the final analysis.

Fractionation Research, Inc. (FRI) considered the use of gamma scan technique for quantitative determination of froth density in high pressure downcomer studies. It was recognized that the lack of accuracy in the downcomer backup models was due to lack of accurate froth density measurement. There were some mean froth density values obtained from visual observation and manometer readings, but there was no downcomer froth density gradient available. Survey of the existing literature related to gamma ray scanning indicated it may be a tool to be used in this context.

DESIGN OF THE EQUIPMENT

The first step in the design of the equipment was to determine the proper gamma ray source that fulfills the froth density measurement requirement. The amount of gamma ray absorbed by different materials is proportional to the product of density and the distance of the material from the source. The weakening of the detected signal due to the spread of rays is inversely proportional to the square of distance from the source. The FRI high pressure column is 1,220 mm (48") and 25 mm (1") thick steel wall. **Figure 1** is the design basis for a 13% downcomer, to get meaningful results one should be able to distinguish a froth density of 0.5 from 0.45 for butane system at pressures up to 34.5 bar (500 psia). The outlet weir length of the 13% downcomer is 952 mm (37.8") with a downcomer width of 230 mm (9"). The scan chord length that divides the downcomer in two equal volume segments, is 782 mm (30.8") and is located 142 mm (5.6") from the column wall. At this point the, the traversing gamma ray beam, ignoring the distances between the source and detector from the column wall, must pass through 150 mm (3") of steel and 782 mm (30.8") of the process fluid. The steel has a density 8 times as much as water, while a 50/50 mixture of butane liquid and vapor at 34.5 bar (500 psia) has a specific gravity of 0.23. Thus the relative absorptivity of the steel wall to that of process fluid is:

$$(2 * 37.5 * 8)/(782 * 0.23) = 3.36$$

Similarly a scan at the column center, has a relative absorptivity of

$$(2 * 25 * 8)/(1220 * 0.23) = 1.43$$

In addition, the signal strength from the scanned downcomer chord to the column center is reduced by:

$$(782)^2/(1220)^2 = 0.41$$

As the scanning proceeds from the column center towards the column wall, in parallel chords, the relative signal strength increases; however, the relative absorptivity of steel wall to the process fluid also increases. The value becomes infinity at the column wall. The determination of the quantitative value of froth density, which varies from point to point under the conditions where over half of the rays are absorbed by the column walls is different from the commercial applications. To accomplish this, a calibration system was developed that scans the column at pre-determined horizontal and vertical positions, once with column full of vapor and a second time with the scanning zone full of liquid, and saving the results as calibration tables. During the test, the measured density at a given location is compared to the calibration value at that position and froth density is calculated. The calibration must be performed with two fluids of known densities that bracket the anticipated process values, for high pressure downcomer studies air and liquid butane were used.

In 1987 a standard 2 Curie Cesium-137 gamma ray source, Texas Nuclear Model 5203, and a scintillation detector, Texas Nuclear Model 9709, were selected. The system included a microprocessor with such features as exponential signal averaging with a variable time constant, automatic source decay compensation, and key or computer input of calibration data. It should be noted that the source strength of 2 Curie is 40 times stronger than the 50 millicurie source used by most commercial scanning operators

POSITIONING MECHANISM

To measure the froth density accurately, one needs to position the source and the detector in the exact location when the empty column calibration scans were created. Previous ultrasonic inspections had shown a variation in wall thickness of about 0.5 mm (0.02") from point to point. In terms of gamma ray absorption, this is equivalent to a 5% variation in froth density. After extensive calculations, a positioning tolerance of 0.25 mm (0.01") was set as a target. It is obvious that such a tolerance was not achievable with suspending the source and the detector from cables, as is the normal practice of the commercial scanners. A tight tolerance provides precise data, but it requires a sophisticated positioning mechanism and sampling time, while a relaxed tolerance results in a less accurate measurement with a mobile positioning mechanism and quicker sampling time. In order to minimize the loss of signal due to spreading, it was decided to place the source and the detector as close as possible to one another, in a horseshoe structure as is shown in **Figure 2**. Vertical positioning within 0.25 mm (0.01") is accomplished remotely from control room utilizing a constant speed motor and a position indicator (displacement transducer). Horizontal positioning, as shown in **Figure 3** is manual, utilizing precision drilled alignment holes on 50 mm (2") increments from the centerline to just inside the column wall. The installation provided a continuous vertical movement of 1118 mm (44") for the aligned source and detector at each of twelve horizontal positions, starting at column centerline.

A geared motor drive moves the horseshoe shaped support plate between two limit switches. A vertical position sensor sends a signal to a vertical position indicator (VPI) in the control room, where the geared motor drive is located. The horizontal position change is accomplished by unbolting both the source and the detector, and setting them in one of the twelve locations and bolting them to the support plate.

Since the source was shielded by using lead container, it weighed 60 kg (130 lb) which was not convenient for manual positioning. To prevent deflection of the horseshoe support plate and make the manual relocation easier, a counterweight was added to the source housing. **Figure 4** is a photograph of the source housing with Cesium-137. To stop emission of radiation at the outer most chord scanning, a lead "doughnut" was added to the front of the source housing.

To increase the vertical scanning range, and resolve the difficulties of horizontal positioning of the heavy Cesium-137 source, the gamma scanning system was replaced with a new system using Cobalt 60 with a half-life of 5.27 years in 2006. The new system has a scanning range of 2921 mm (115"), compared to 1118 mm (44") of the previous setup. The strength of the Cobalt-60 is 500 mCi, **Fig 5** is photograph of the side view of the new source.

The horizontal movement is accomplished by a sliding plate with pegs located in 50 mm (2") intervals. **Figure 6** shows the detector and the horizontal sliding plate. The detector is connected to a LB444 device from Berthold Technologies[3]that sends the signal to a PC based software. The vertical position mechanism was also replaced with a movable vertical mast and a chain attached to a hoist from the deck above and is shown in **Figure 7**. A Celesco strain gauge provides the signal to the positioning PC indicating the vertical location of the source and the detector.

DOWNCOMER STUDIES

One of the objectives of these studies was to obtain froth density profile in downcomers approaching flood at high pressure. Specifically performance of a straight 13% downcomer and a sloped downcomer with 18% of the column area at top and 8% at the bottom. The system used was iC_4/nC_4 at pressures ranging from 11.4 to 34.5 bars (165 to 500 psia).

Figure 8 is a column schematic of the gamma scan region. Six trays were installed on 610 mm (24") tray spacing. Each tray had a bubbling area 74% of the column cross-sectional area with 12.7 mm (0.5") holes on 33 mm (1 5/16") triangular spacing and a 940 mm (37") long and 50 mm (2") high outlet weir. The straight downcomer had an area equivalent to 13% of the column cross-sectional area and a downcomer clearance of 38 mm (1.5"). The traversing mechanism provided a continuous vertical movement of 1118 mm (44") for the aligned source and detector at each of the twelve horizontal positions spaced 50 mm (2") apart, starting at the column center line. The trays were installed in such a way that the bottom of the Tray 4 was aligned with the lowest point of the traversing mechanism and the downcomer from Tray 5 was in the half of the column that the traversing mechanism scanned..

Calibration was done in two parts. First, the column was filled with air or the process vapor and the calibration reference ratio was recorded at the selected horizontal and vertical locations. Second, the column was filled with liquid butane of known composition and the process was repeated for the same points. For each measurement, the temperature was recorded enabling the proper calculation of the vapor and the liquid density associated with the measured signal. These calibration reference ratios and densities were saved in a file, which was available to the densitometer prior to measuring the density under the actual process condition. After measuring the process density, ρ, the liquid volume fraction at process condition, φ, can be calculated using the vapor density, ρ_V, and the liquid density, ρ_L, from

$$\varphi = (\rho - \rho_V)/(\rho_L - \rho_V)$$

Figure 9 shows a plot of liquid volume fraction in downcomer as a function of elevation for three horizontal scanning positions for the 11.4 bar (165 psia) butane system at 95% of flood. There appears to be a consistent radial density gradient. Action at the mouth of the downcomer with decay of the froth immediately above it is particularly evident. The same gradient trend is observed for the same system at 34.5 bar (500 psia) and 95% of total reflux flood, as shown in **Figure 10**.

The liquid volume fractions for vertical positions between 711 and 1118 mm (28" and 44") match very closely above the bubbling area for all the horizontal positions scanned for the butane system at total reflux and 95% of flood and 11.4 bar (165 psia), **Figure 11**, and 35.4 bar (500 psia), **Figure 12**.

To observe the effect of sloping the downcomer on froth density, trays with 12.7 mm (0.5") holes, a hole area 10.6% of the bubbling area, a 50 mm (2") outlet weir, and sloped downcomers with areas at top and bottom equal to 18.9 % and 8.8% of the column cross-sectional area were utilized. **Figure 13** shows a plot of liquid volume fraction as a function of elevation in downcomer at 500 mm (20") horizontal position from the tray center for butane system at 20.7 bar (300 psia) pressure approaching the flood conditions at total reflux. The straight downcomer shows much higher liquid fraction throughout the downcomer, and has a significant buildup of relatively high liquid fraction froth over the downcomer. Also, the liquid off Tray 5 (610 mm (24") elevation) has a lower liquid fraction for the sloped downcomer. Integration of the froth density in the downcomer as a function of height results in the equivalent clear liquid depth. The comparison of the results obtained by gamma ray densitometer on Tray 4 with bubblers results obtained on Tray 3, indicated this method can be used to obtain clear liquid on trays[4].

TRAY HOLDUP STUDIES

Having proved the value of gamma ray densitometer in the downcomer studies, it was decided to use the method for tray holdup studies. Liquid holdup on a tray depends on the vapor rate, liquid rate, and system. It increases with increasing liquid rate, but decreases with increasing vapor rate. Although some of the existing correlations predict the general trend of the holdup fairly well, none of the correlations predicts holdup adequately. Since the scanning equipment can only scan half of the column, the readings from the outlet half of Tray 5 were combined with the readings of the inlet half of Tray 4 to construct contour plot that approximates the density profile of the composite tray.

The liquid holdup studies were conducted for cyclohexane at 1.65 bar (24 psia) on an 8 percent hole area sieve tray with 12.7 mm (0.5") holes on 38 mm (1.5") triangular centers. Outlet weir was 50 mm (2") and 150 mm (6") high and 940 mm (37") long. Tray spacing was set to 610 mm (24"). Froth densities on the tray were measured by using both the gamma-ray densitometer and hydraulic bubblers over a range of vapor and liquid rates at several horizontal and vertical positions between trays. To relate the measurements to the existing data, the composition effect was eliminated by using pure Cyclohexane rather than a mixture of Cyclohexane/n-Heptane. **Figures 14** and **15** are typical results for this system with 50 mm (2") and 150 mm (6") weir height respectively. Plotting contours of constant froth density lines for the same process conditions, one may conclude, the froth densities (and therefore the holdup) above the weir are essentially the same for both weir heights. Thus a hold up model should be compromised of two elements, the first is from the tray floor to the weir top that has to be a function of the weir height; and the second element is above the weir and is independent from the weir height.

GAMMA SCANNING IN A PACKED COLUMN

FRI tested a high capacity proprietary random packing (No. 2.5 Nutter Ring), using Mobil Pour Pan Distributor designed for high liquid rates and the FRI tubed drip pan distributor for low rates, with butane systems. The tests were conducted at pressures ranging from 6.89 to 27.6 bar (100-400 psia).

Figure 16 shows the column configuration for these tests. The packing was a high open area (98% void) stainless steel, which leaves 2% additional steel that the gamma rays have to penetrate. At column center, the column thickness that the rays traverse is 50 mm (2"); the packing contribution is

Packing Thickness Contribution = 1220 X 0.02 = 25 mm (1")

This indicates, in case of packed column the beam must penetrate 50% more steel than the empty column. In addition, for trays the tests focus was on the region where the process fluid was about 50% liquid. Packed columns normally operate in the range of 1-20% of liquid holdup[5]. **Figure 17** display the liquid volume fraction at the column center at an elevation of 1370 mm (54") above the support plate for operating pressures 6.89 to 27.6 bar (100-400 psia) for total reflux runs. Previously, liquid holdup in a randomly packed bed below loading point was found to be dependent on liquid parameters only and the effect of vapor rate was regarded as minimal[6,7]. These conclusions were based on studies usually with atmospheric air water system. **Figure 18** for the constant liquid rate data at 20.7 bar (300 psia) and butane system using the Nutter rings shows the vapor rate effects the liquid holdup.

Since 1987 FRI has used gamma scanning technique to measure the liquid holdup in random and structured packing. **Figure 19** shows a typical plot of liquid holdup as a function of percent flood at total reflux operation for runs with iC_4/nC_4 at 11.4 bar(165 psia) and 20.7 bar (300 psia) and C_6/nC_7 at 1.65 bar (24 psia). The liquid volume fraction increases with increasing rates. With iC_4/nC_4, the liquid volume fraction increased significantly as the percent flood increased from 60% to 90%. For C_6/nC_7 system, there was almost a linear relationship between the liquid holdup and percent flood.

CONCLUSIONS

These results indicate that with sufficiently powerful gamma ray source, and a mechanism that can align the source and the detector along the scan cord accurately, one can make quantitative measurements of froth density and liquid holdup in downcomers and packed bed. For accurate measurements, one need to calibrate the column with two fluids with densities that bracket the process fluid during the operation. Meaningful results are obtained only if the vertical and horizontal positions of the source and the detector are reproducible during calibration and the operation.

Nomenclature

iC_4 = Iso-butane
nC_4 = Normal Butane
C_6 = Cyclohexane
nC_7 = Normal Heptane
F_s = F factor, $m/s(kg/m^3)^{1/2}$

Greek Letters

ρ = Process density, (kg/m3)

ρ_L = Liquid density, (kg/m3)

ρ_V = Vapor density, (kg/m3)

φ = Liquid volume fraction

REFRENCES

1. Beckmann, R.B. and Hwa, C.S., September 1960, "Radiological Study of Liquid Holdup and Flow Distribution in Packed Gas-Absorption Columns", AIChE Journal, 6(3):359.
2. Severance, W.A.N., DuPont Co., September 1981, "Advances in Radiation Scanning of Distillation Columns", CEP, 81(9), P.38, 1981.
3. Berthold Technologies USA, LLC, 99 Midway Lane, Oak Ridge, TN 37830
4. Kunesh, J.G., King, D.W., and Fitz, C.W., 1992, "Use of Gamma Scanning to Obtain Quantitative Hydraulic Data on a Commercial Scale", Distillation and Absorption 92.
5. Bilet, R. and J. Mackowiak, 1982, "How to Use Absorption Data for Design and Scale-up of Packed Columns", EFCE Working Party on Distillation, Absorption and Extraction", Helsinki Meeting, June 2-4, 1982.
6. Bemer, G.G. and Kalis, G.A.J., Trans. I Chem. E., Vol. 56, p. 200(1973).
7. Buchnan, J.E., Ind. Eng. Chem. Fundamentals, Vol. 6, p.400 (1967).

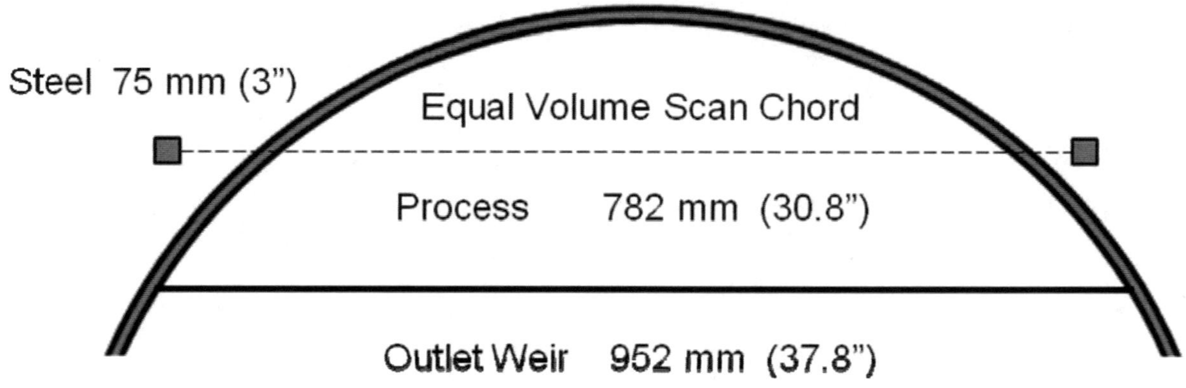

Figure 1. Design Basis

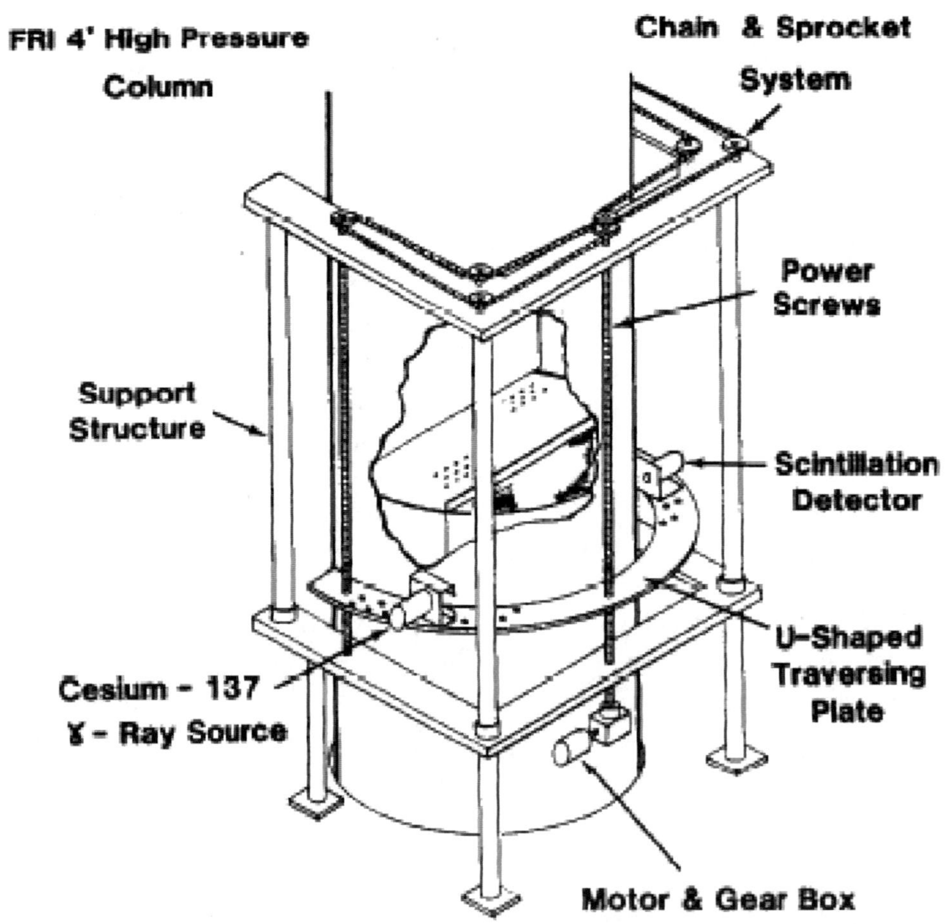

Figure 2. Vertical Positioning Mechanism

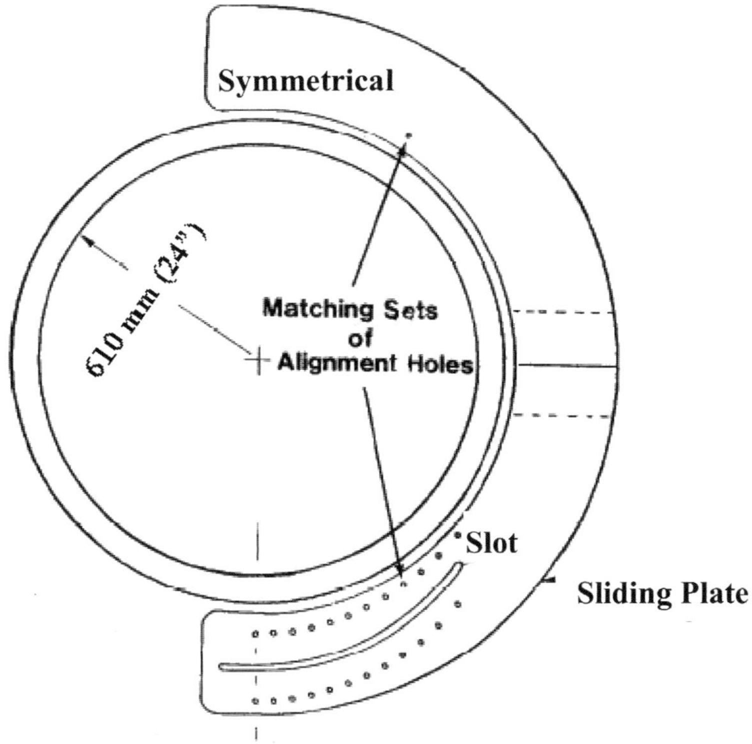

Figure 3. Horizontal Positioning Design

Figure 4. Cesium-137 Source Housing

Figure 5. Side View of Cobalt-60 Source

Figure 6. Detector and Horizontal Positioning Sliding Plate

Figure 7. New Vertical Positioning System

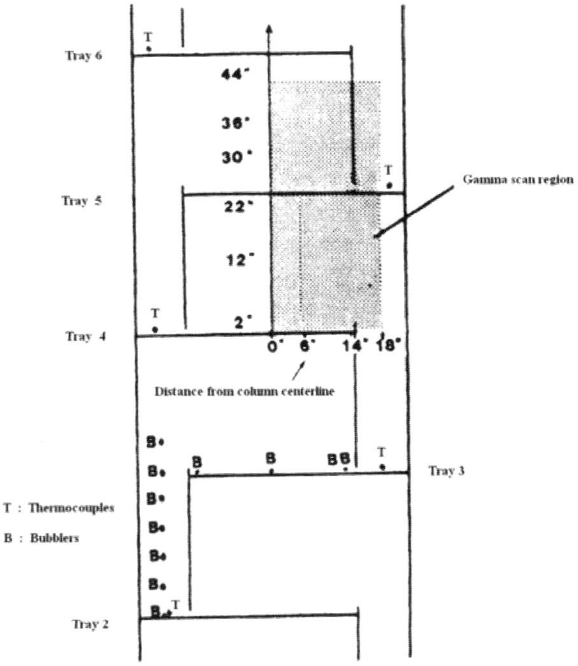

Figure 8 Column Schematic Side View

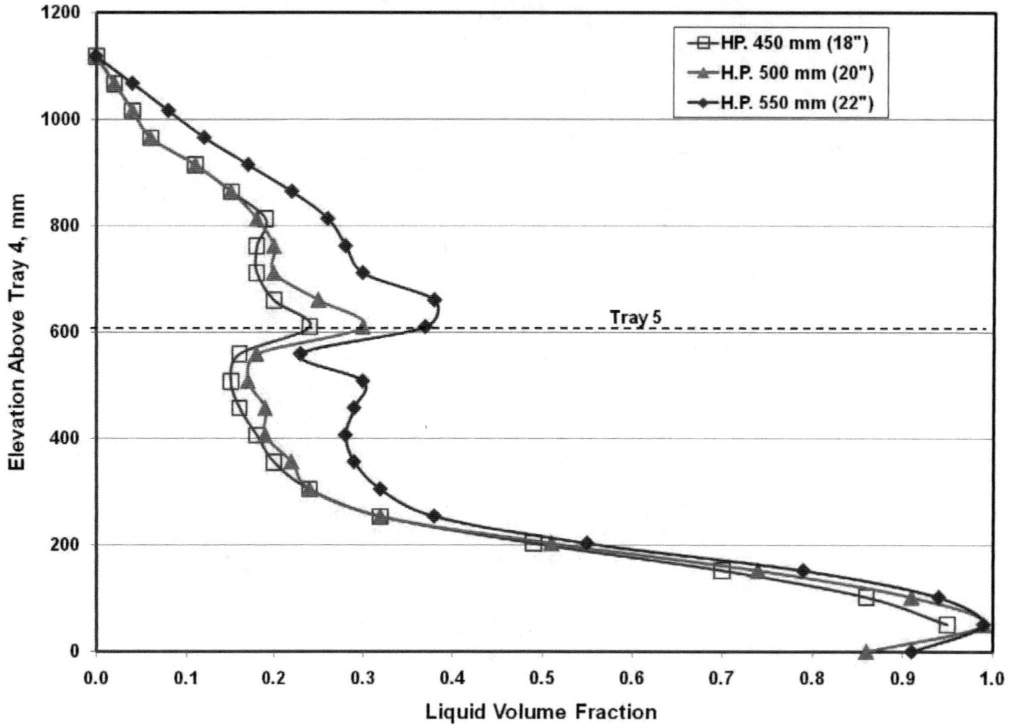

Figure 9. Downcomer Scan (Straight Downcomer)
iC_4/nC_4, 11.4 bar (165 psia), 95 % of Flood

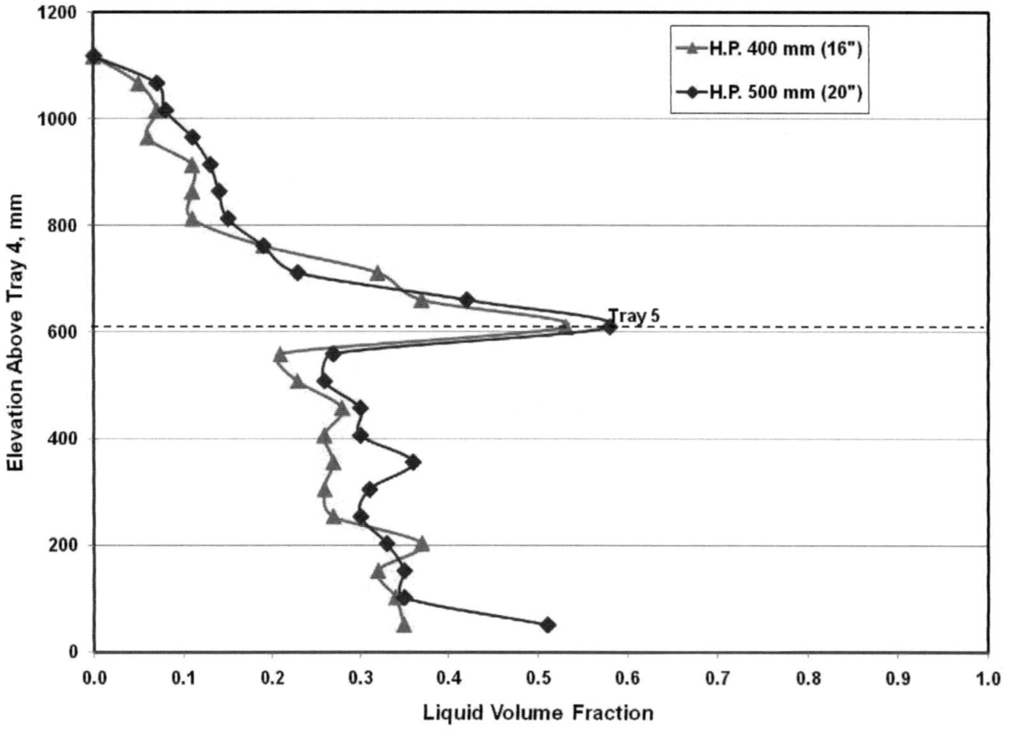

Figure 10. Downcomer Scan (Straight Downcomer)
iC_4/nC_4, 34.5 bar (500 psia), 95 % of Flood

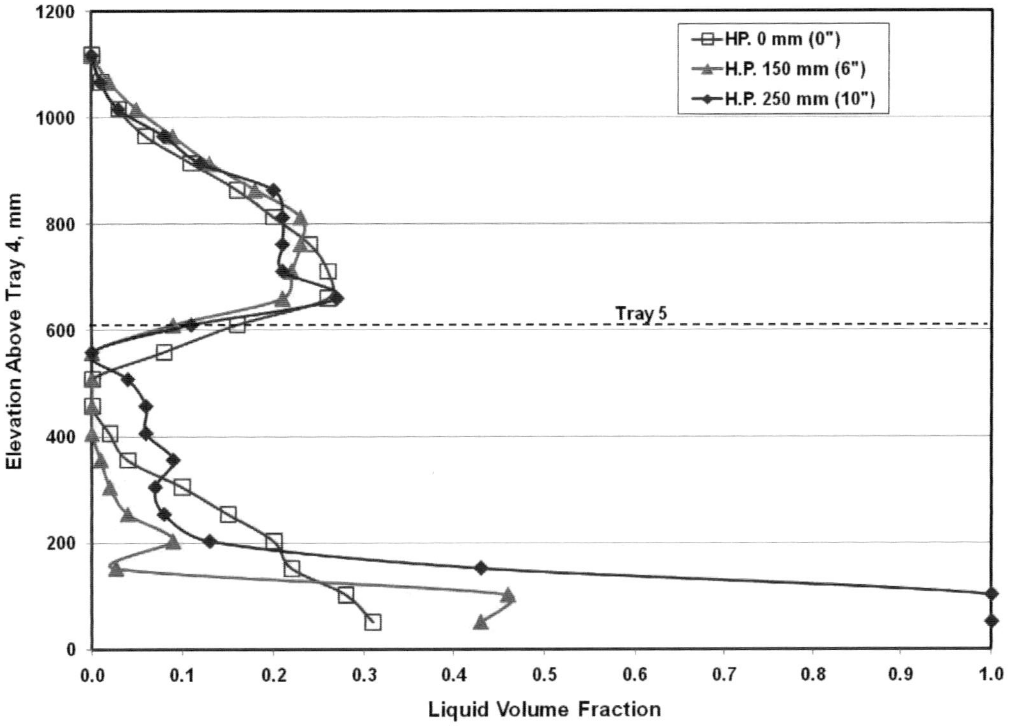

Figure 11. Tray Scan (Straight Downcomer)
iC_4/nC_4, 11.4 bar (165 psia), 95 % of Flood

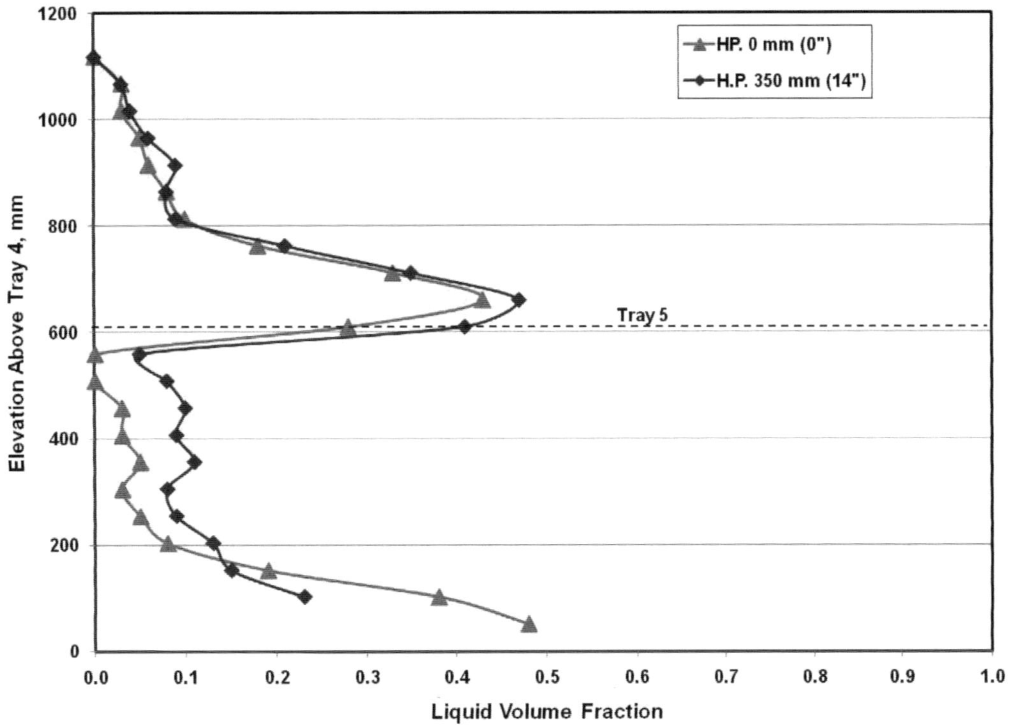

Figure 12. Downcomer Scan (Straight Downcomer)
iC_4/nC_4, 34.5 bar (500 psia), 95 % of Flood

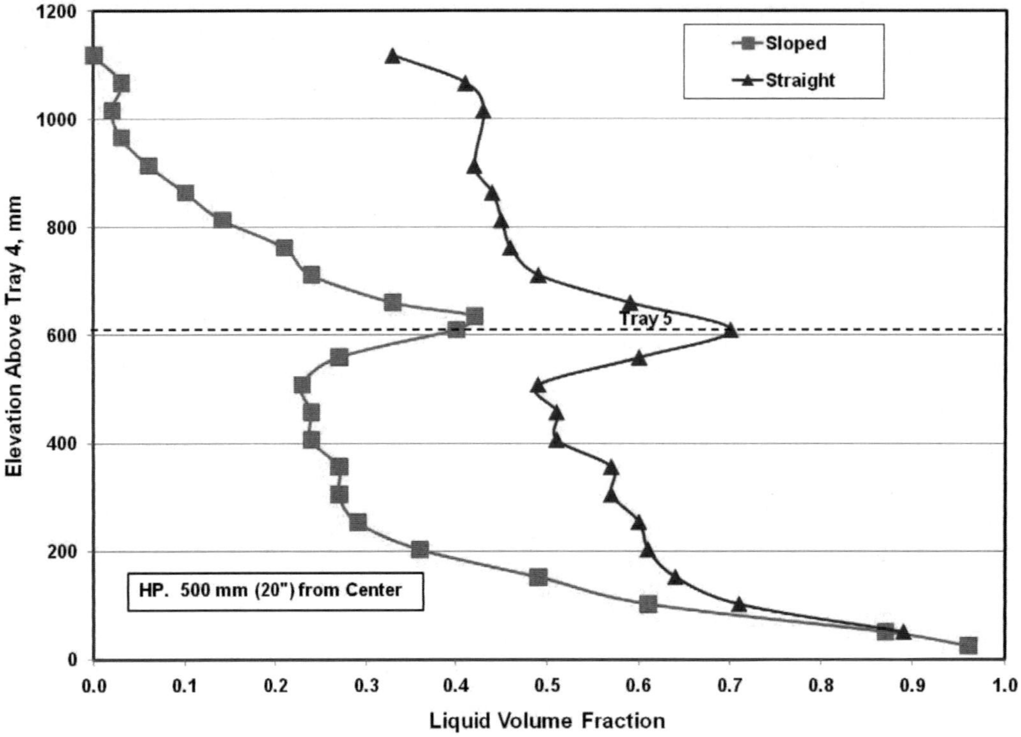

Figure 13. Straight vs. Sloped Downcomer
iC_4/nC_4, 20.7 bar (300 psia), 95 % of Flood

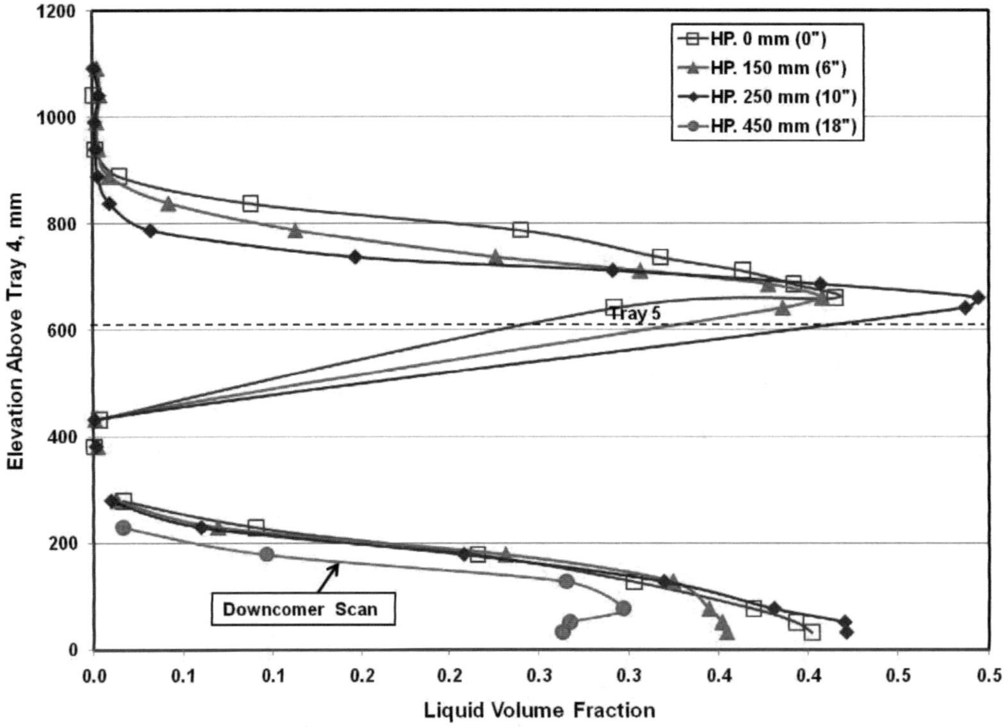

Figure 14. Tray Holdup Scans, 50 mm (2") Weir Height
Cyclohexane, 1.65 bar (24 psia)

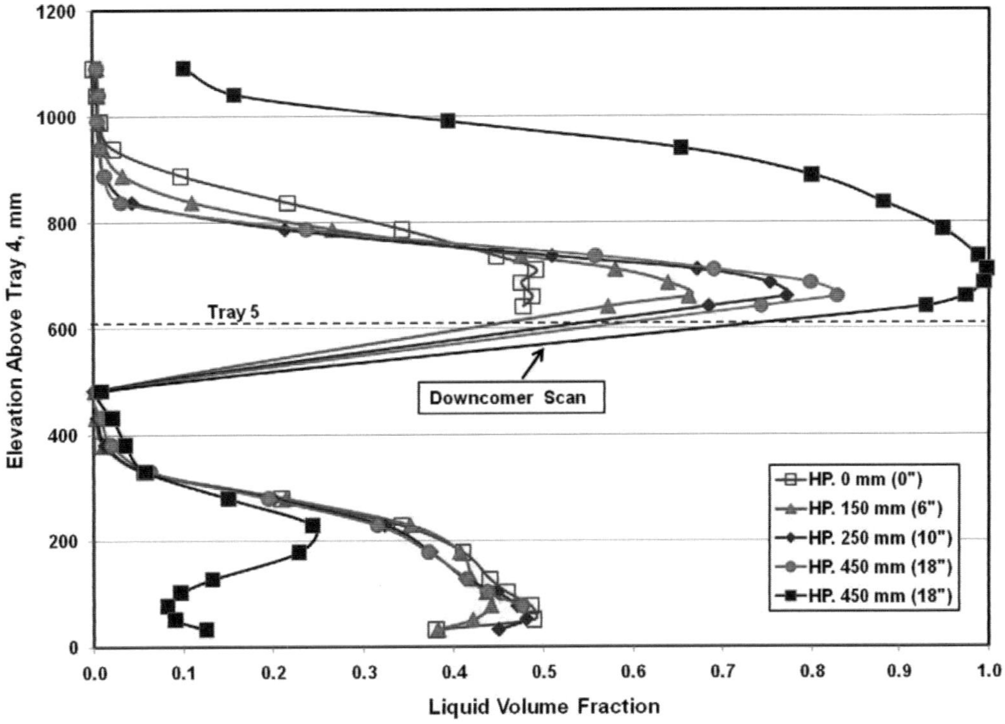

Figure 15. Tray Holdup Scans, 150 mm (6") Weir Height Cyclohexane, 1.65 bar (24 psia)

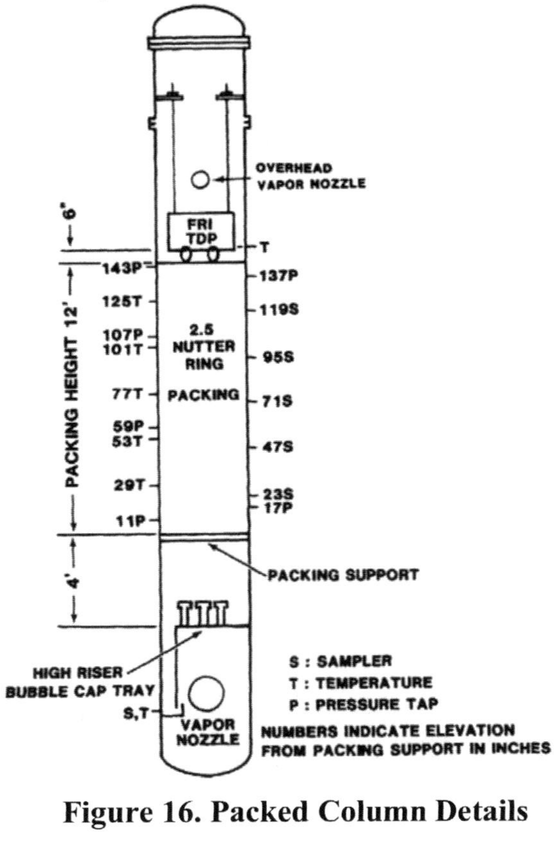

Figure 16. Packed Column Details

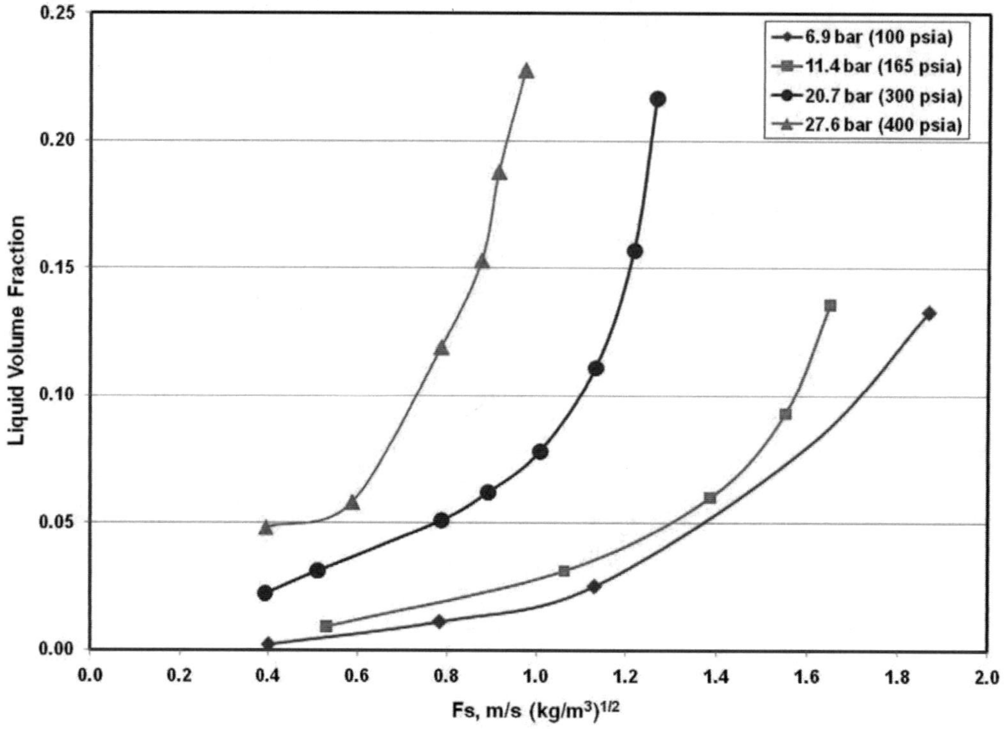

Fig 17. Gamma Ray Scans of No. 2.5 Nutter Ring iC_4/nC_4 System, Total Reflux Runs (1370 mm Elevation)

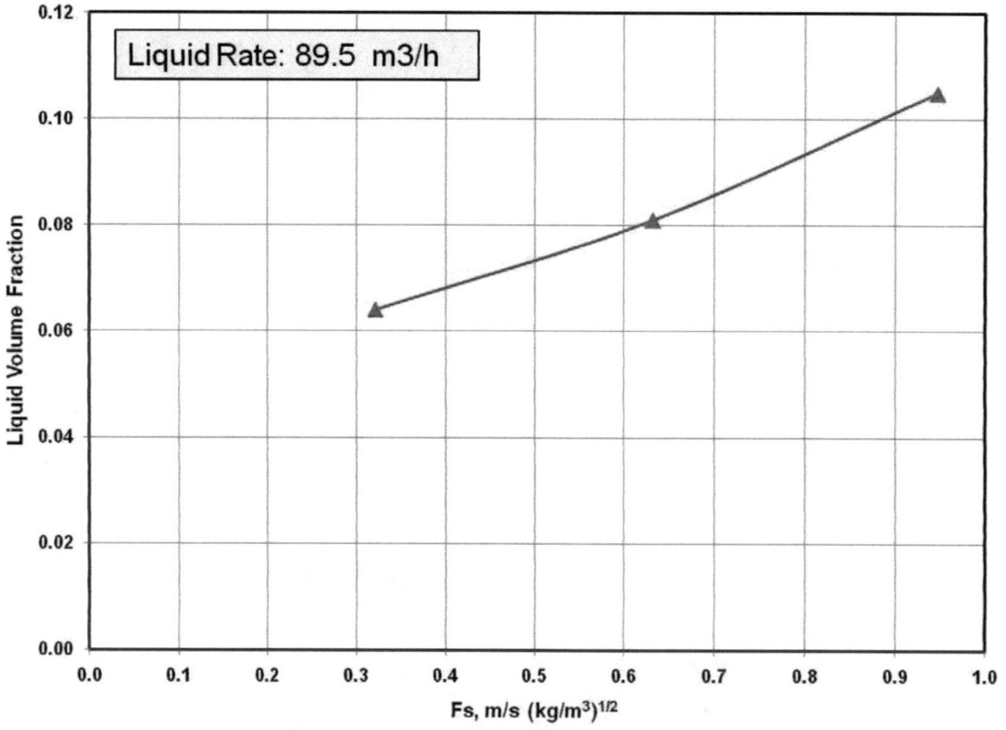

Figure 18. Gamma Ray Scanning of No. 2.5 Nutter Ring iC_4/nC_4 System, Constant Liquid Runs (1370 mm Elevation)

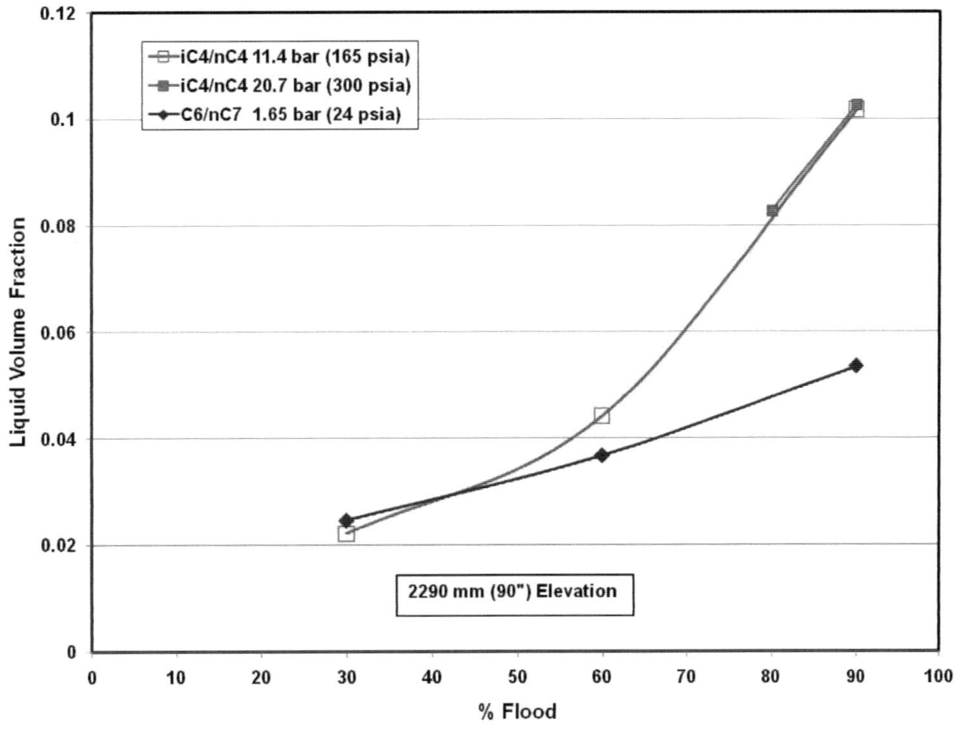

Figure 19. Structured Packing Liquid Holdup Koch-Glitsch Intalox®2T

Kettle Troubleshooting

By

Henry Z. Kister, Fluor,
Aliso Viejo, CA, USA

and

Milton A. Chaves, Hess Corporation,
Tioga Gas Plant, Tioga, ND

Paper 84a
Presented at the Distillation Topical Conference,
AIChE Spring Meeting, San Antonio, Texas, March 23, 2010

Cover Story

Kettle Troubleshooting

Henry Z. Kister
Fluor Corp.
and
Milton A. Chaves
Hess Corp.

Here's proof that kettle reboilers can behave like thermosiphons and thereby bottleneck an entire plant. Understand the mechanism to blame and avoid it with these prevention and troubleshooting tips

A survey of tower failures [*1*] ranks kettle reboilers as the most troublesome reboiler type in the chemical process industries (CPI). Excessive pressure drop in kettle reboiler circuits is the prominent kettle malfunction, causing liquid to back up in the tower base beyond the reboiler return elevation. This high liquid level leads to premature flooding and capacity loss. In almost all the surveyed cases, the excess pressure drop is due to hydraulic restrictions in the kettle inlet or outlet piping or at the kettle entrance [*2*].

What is often unappreciated is that in many of the troublesome cases, entrainment plays a key role in tower base backup. Entrainment from kettle reboilers incurs a static head and more friction, which exacerbates kettle outlet-line pressure drop [*3*]. Furthermore, the entrained liquid drops get knocked out in the tower and return to the kettle, raising the flows and pressure drops through the kettle circuit.

Entrainment from kettle reboilers occurs when the disengagement space above the bundle is insufficient to disentrain liquid droplets. Several criteria and rules of thumb for preventing entrainment are available, the most popular of which is keeping the top tube row not higher than 60% of the shell diameter [*4–6*]. Other criteria include keeping a minimum height of 12 in., or 1.3–1.6 times the bundle diameter (whichever is greater) above the liquid level [*7*]. A more rigorous criterion for setting the disengagement space is proposed by Tammami [*8*].

The problem for troubleshooters is that until now, there have been limited literature reports of entrainment from kettle reboilers. Hartman and Hanson [*9*] reported a case with three parallel, dissimilar kettles, where an excessive liquid level set in one draw compartment induced overflows in all-three draw compartments, entraining liquid into the vapor lines, high pressure drops in the kettle circuits, and flooding in the tower.

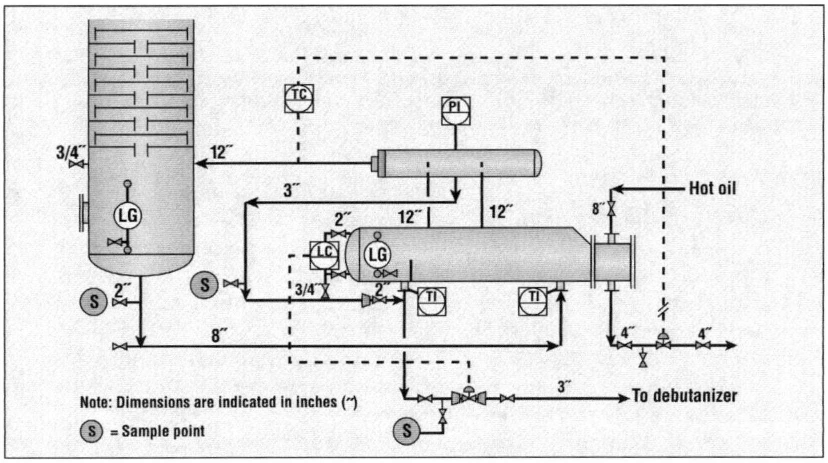

FIGURE 1. Comprehensive field tests, conducted on the depropanizer kettle reboiler circuit in the Hess lean-oil gas plant, verified that kettles can thermosiphon

Rubbers et al. [*3*] reported entrainment from a kettle reboiler in extractive distillation service, leading to excessive pressure drop and to an overflow of the chimney tray in the tower that was feeding the kettle. Rubbers' field investigation, which utilized surface temperature surveys, pressure drop measurements and hydraulic calculations, identified boiling maldistribution in the reboiler as the root cause. A theory explaining all the observations subdivided the reboiler into a stagnant region, an intense boiling region, and a kettle region. The intense boiling region was characterized by entrainment into the overhead line, low fractional vaporization, and relatively high concentration of light components.

This article shows that this intense boiling region is not unique to extractive distillation but may occur in many other applications. Further, we demonstrate that the intense boiling action can develop into a full-blown thermosiphon action. A thermosiphon reboiler is one in which substantial liquid circulation develops between the reboiler and the column due to the density difference between the liquid in the column and the two-phase material in the vapor return line.

Although one of the authors encountered cases where a kettle reboiler was hypothesized to behave like a thermosiphon reboiler, this is the first time where this behavior was actually monitored and demonstrated by comprehensive field tests. The findings demonstrate that entrainment from kettles is a major issue that needs to be considered when kettle reboilers are designed, revamped or troubleshot. Meanwhile, the methods described here illustrate how to best troubleshoot a kettle reboiler and diagnose entrainment.

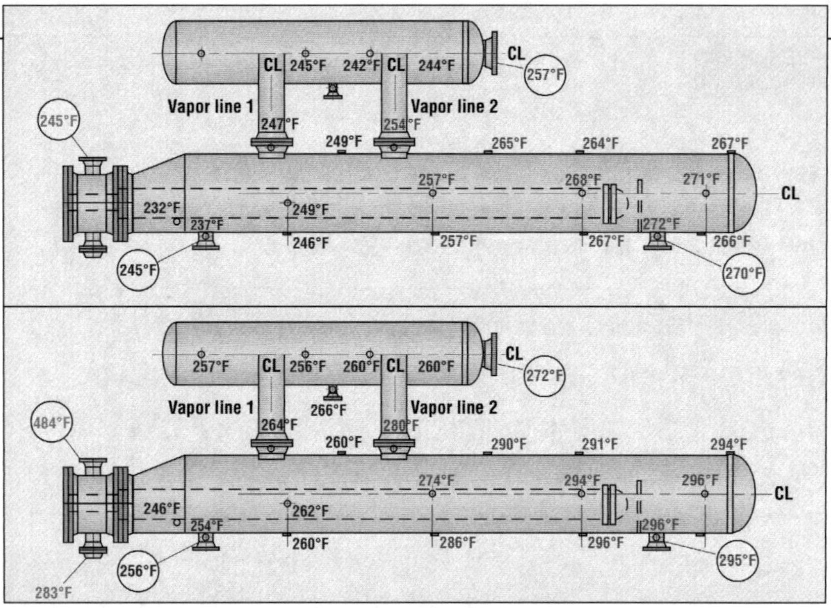

FIGURE 2. Surface temperature surveys show an inlet region, with temperatures close to the kettle liquid feed, and a much hotter outlet region. Test a (above) is at 140 gal/min feed; Test b (below) is at 195 gal/min feed

Operating history

The depropanizer in the Hess lean-oil gas plant in Tioga, N.Dak. contains 40 two-pass trays at 24-in. spacing and is equipped with a kettle reboiler heated by lean oil, which enters the reboiler at 490°F and leaves at 285°F (Figure 1). Process liquid from the depropanizer bottom enters the kettle via an 8-in. line. There are two 12-in. dia. vapor-lines leaving the reboiler. These lines go into a disengagement drum that has a 30-in. inside diameter (ID) and is 12-ft long. Vapor from the drum returns to the depropanizer via a horizontal, 12-in. dia. line. Liquid knocked out in the drum flows via a 3-in. dia. line into the kettle liquid outlet. Inside the kettle, unboiled liquid overflows a baffle into the kettle draw compartment and is withdrawn on level control. The kettle shell is 36 $^5/_8$-in. ID and contains 344 tubes that are $^3/_4$-in. outside diameter (OD) and 20-ft long, set on a 1-in. square pitch and arranged in four tube passes. This design gives a total heat transfer surface area of 1,350 ft^2. The overflow baffle is 10 $^5/_{16}$-in. below the top of the shell.

The tower and reboiler system have the nameplate capacity of 263 hot gal/min of feed. For decades, the plant operated well below its nameplate capacity and the tower experienced no bottleneck. Recently, economics favored higher plant throughput. When the depropanizer feed was raised to 205 gal/min, flooding occurred near the bottom of the tower. This flood bottleneck limited throughput of the entire gas plant.

Operating experience showed that at feedrates of less than 160 gal/min, the column operated well, and no level was seen in the tower's bottom level glass. As the feedrate was raised, a liquid level became visible. As feedrates were raised from 160 to 205 gal/min, this level was observed to first rise very slowly, then faster. Above 205 gal/min it shot right up and above the reboiler return inlet. Once it rose above the reboiler return inlet, the tower was observed to flood.

A preliminary hydraulic evaluation and tower gamma scans confirmed that the flooding initiated at the reboiler circuit and not in the tower. The force balance for the reboiler circuit showed that at the last feedrate below flooding (200 gal/min), the pressure drops in the inlet and outlet lines were too small to cause anywhere near the observed liquid stackup in the tower bottom sump. This preliminary force-balance calculation did not account for liquid entrainment from the kettle.

Field testing

Theories about the cause of the bottleneck included entrainment from the kettle, plugging of the reboiler inlet line, heavies accumulation in the kettle and more. To narrow down the possible causes, it was decided to perform a two-day test at two depropanizer feedrates: 140 gal/min, where little liquid was observed at the bottom of the tower; and 195 gal/min, about the highest feedrate at which the tower operated steadily. Fluctuations in the tower feed and operation were minimized during each test.

A complete set of tower gamma scans was performed at the higher tower feedrate of 195 gal/min. It showed that all the trays were in place, holding 3–7 in. of liquid, with center downcomers holding 6–9 in. of liquid. Little entrainment was visible in the vapor spaces, and there were no signs of flooding or even an approach to flooding in the tower. This was confirmed by the low pressure drop in the tower.

At each feedrate, the reboiler inlet line was gamma scanned at seven locations to look for possible blockage. Measured liquid densities at all points and planes did not vary. The hot specific gravity measured by the scans (0.46) was incredibly close to that calculated from the temperature and composition of the stream (0.45). This test showed that the reboiler inlet line was liquid-full and that there was no indication of any blockage in it.

Neutron backscatter measurements of the liquid level in the tower sump were in excellent agreement with the tower-base level glass. At the 140 gal/min test, the liquid level in the tower was 4 in. below the bottom tangent line (*TL*), rising to 14 in. above the *TL* in the 195 gal/min test.

The two vapor lines from the reboiler to the disengagement drum, and the horizontal vapor line from the drum to the tower, were gamma scanned along various planes. These scans established that there was a large amount of entrainment from the kettle reboiler to the disengagement drum and from the drum to the tower. This entrainment increased with the tower feedrate.

Several extensive tests followed, and are described in detail in the box that begins on p. 28. Summarized below are the key findings.

Laboratory analysis. Lab analyses found that at the 140 gal/min feedrate test, the disengagement drum liquid was only slightly heavier than the depropanizer bottoms feeding the kettle, indicating that the drum liquid came from the reboiler inlet region. The converse occurred in the

TABLE 1. CHEMICAL ANALYSIS OF REBOILER INLET & OUTLET STREAMS, WT.%

	140 gal/min test			195 gal/min test		
Component	DC_3* bottoms	Drum liquid	C_4+	DC_3* bottoms	Drum liquid	C_4+
C_3	0.38	0.27	0.08	0.31	0.31	0.21
C_4	71.21	67.29	47.63	65.44	52.79	46.02
C_5	17.85	19.55	24.58	21.21	24.47	25.37
C_6	5.81	4.47	13.46	2.39	10.21	11.17
C_7	2.82	4.55	8.00	2.01	9.71	11.49
C_8	1.52	3.50	5.50	4.98	2.04	5.33
C_9+	0.41	0.38	0.76	3.64	0.48	0.39

* Depropanizer

TESTING METHODS

Lab analyses

The depropanizer bottoms, the disengagement drum liquid, and the C_4+ (kettle draw liquid stream combined with disengagement drum liquid) were sampled and analyzed. The sample points are marked on Figure 1. The C_4+ composition was verified by a component balance on the downstream debutanizer.

The results are shown in Table 1. At the 140 gal/min feedrate test, the disengagement drum liquid composition was only slightly heavier than the depropanizer (DC_3) bottoms feeding the kettle, and much lighter than the C_4+ stream. This indicates that the drum liquid for this test was entrained from the reboiler inlet region, where the depropanizer bottoms stream enters the kettle. The converse occurred in the 195 gal/min feedrate test, where the disengagement drum composition was only slightly lighter than the C_4+ stream, and much heavier than that of the depropanizer bottoms stream. The reason for this behavior is discussed later (Composition and temperature behavior, p. 31).

Temperature surveys

Surface temperatures were measured using a laser-guided pyrometer, which was calibrated with measurements of the depropanizer bottoms and the kettle liquid draw temperatures. Figures 2a and 2b show the measurements for the 140 and 195 gal/min feed tests, respectively. Pyrometer measurement locations are drawn roughly to scale on Figures 2a and 2b, with the measured temperature next to each. Surface temperatures were measured at the top, bottom, and sides of the kettle, and in the liquid disengagement drum. For clarity, only the east side temperatures are shown. Those measured on the west followed the same trend. Note that the surface temperatures were measured near the wall, at a varying distance from the tube bundle. The temperatures of the streams that enter and leave the system were measured by the process thermocouples and are the encircled temperatures marked on Figure 2.

Figures 2a and 2b distinguish two regions along the length of the kettle. In the inlet region, that stretched from the channel head to Vapor line 2 (the second nozzle from the channel head), the temperatures were only slightly higher (by no more than 4–6°F) than the kettle inlet temperatures. In the second region, which stretched from Vapor line 2 to the liquid draw, the kettle temperatures began rising right after Vapor line 2, finally reaching the liquid draw temperatures.

Temperatures in the disengagement drum were much the same as the kettle liquid inlet temperatures in the 140 gal/min test, and slightly (about 4°F) higher than the kettle liquid inlet temperature in the 195 gal/min test. Temperatures in Vapor line 1 were colder than in Vapor line 2. This supports earlier comments that much of the liquid that reached the disengagement drum was entrained from the inlet region of the kettle.

There were differences between the temperature surveys for the two tests. The temperatures in the 140 gal/min test were lower than those in the 195 gal/min test, by about 11°F at the kettle inlet and by about 25°F at the kettle liquid draw. Also, the temperature difference between kettle liquid outlet and inlet rose from 25°F in the 140 gal/min test to 39°F in the 195 gal/min test. Finally, while the kettle surface temperatures measured were completely steady and repeatable in the 140 gal/min test, they varied, by 2–3°F at the cold end, and by as much as 8°F near the hot end, in the 195 gal/min test. The reason for these differences is discussed later.

Reboiler neutron backscatter scans

This technique shoots neutrons about 6 in. into a vessel. Neutrons are reflected upon collision with hydrogen nuclei. The reflection (backscatter) count provides a measure of the concentration of hydrogen nuclei near the vessel wall. The count is high for liquid hydrocarbons where hydrogen-containing molecules are close together, and low in vapors where molecules are far apart. So a high count indicates liquid, while a low count signifies vapor. These scans only measure the liquid content near the wall and not in the bulk of the vessel.

Six neutron-backscatter scans were performed in each test. Figure 3 shows the scan locations and the results for the 195 gal/min feedrate test. The results for the 140 gal/min test were similar. In each scan, neutrons were shot every 2 in. circumferentially from the kettle top to bottom. Figure 3 plots the detector count (*x*-axis) against distance along the circumference from top to bottom.

Figure 3 shows a large amount of liquid in the vapor space above the kettle overflow weir near the kettle inlet. Scanlines 1 to 3 show high liquid content 10 in. (circumferentially) above the weir, much the same as (in Scanline 1, even higher than) the liquid content below the weir. Scanline 2 in the reboiler inlet region showed significant holdup of liquid in the vapor space all the way to the top of the kettle. After vapor nozzle 2, the liquid content in the vapor spaces gradually declined along the kettle length.

Below the weir, most scanlines showed a fairly uniform liquid content. The liquid content was higher for Scanlines 5 and 6, especially lower in the exchanger, indicating less frothy liquid. The liquid content below the weir was least in the inlet region.

The liquid disengagement drum was also neutron-scanned, with three different scanlines circumferentially from the top to the bottom. The drum did not hold much of a liquid level, less than about 2 in.

Tracer injection tests

Non-volatile tracer was injected into the reboiler feedline during the

195 gal/min feedrate test, where the drum liquid was much heavier than that of the depropanizer bottoms. The reason for this result is discussed in the section Composition and temperature behavior (box, p. 31).

Temperature surveys. Surface temperature surveys found that there was a clear split of the kettle reboiler into an inlet region, with temperatures not much warmer than the liquid feed, and an outlet region, where most of the temperatures rise occurred.

Reboiler neutron backscatter scans. Reboiler neutron backscatter scans showed heavy entrainment in the inlet-region vapor space, between the kettle liquid inlet and Vapor line 2, gradually declining as one moves from Vapor line 2 toward the liquid draw nozzle.

Tracer injection tests. Tracer injection tests provided conclusive evidence that entrainment from the kettle was large and came primarily from the reboiler inlet area. The tests determined that the rate of liquid entrainment from the kettle to the drum was 62 gal/min. This rate was independently confirmed by measuring the column base filling rate once the drum drain valve was shut, and from bubble point calculations.

Kettle force balances

Kettle circuit force balances (Table 2) were applied to determine the aeration of the kettle liquid and the entrainment rate from the reboiler and disengagement drum to the tower. Figure 5 illustrates the force balance for the 195 gal/min test. The force bal-

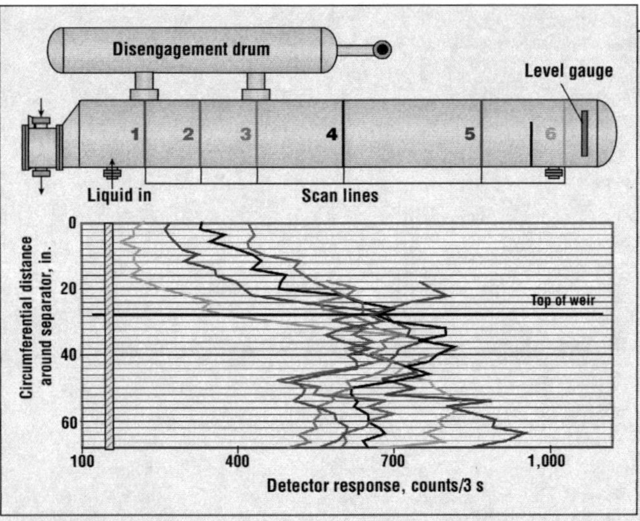

FIGURE 3. Neutron backscatter scan of the kettle reboiler at 195 gal/min feed shows heavy entrainment in the inlet region vapor space, gradually declining from Vapor line 2 toward the liquid draw nozzle

195 gal/min feedrate test. Detectors were mounted at the feedline just before reboiler entry, at the two reboiler vapor outlet lines to the disengagement drum, at the horizontal vapor outlet line from the drum to the tower, and at the reboiler liquid outlet line.

Figure 4a shows the detector responses to a pulse injection. Tracer injected at time zero reached the reboiler inlet 80 s later, giving a sharp peak (black curve), followed immediately (10 s later) by a sharp peak at Vapor line 1 (red curve). The tracer did not finish entering the reboiler when it was detected at Vapor line 1. Tracer began coming out of Vapor line 2 (blue curve) about the same time when it peaked in Vapor line 1, peaking at about 140 s. The peak was lower in counts but of a wider area, suggesting that the tracer was reasonably evenly split between Vapor lines 1 and 2. Tracer started exiting the drum Vapor line (green) at about 100 s, peaking at about 140 s.

The detection of tracer in the disengagement drum and its vapor outlet line provides conclusive evidence for entrainment. The sharp peaks, and the short duration from tracer entry to the peaks in the kettle vapor outlet lines, provide conclusive evidence that the entrainment was large and occurred primarily in the reboiler inlet region. Tracer entering the reboiler was immediately sent overhead via Vapor line 1, and after a short duration also via Vapor line 2.

The tracer reached the liquid outlet (pink curve) about 190 s after injection, and peaked at 450 s, leveling off after that.

Can the entrainment be measured?

In an endeavor to measure entrainment rate from the disengagement drum, the tracer injection test was repeated, but this time the valve in the 3 in. line draining liquid from the drum to the kettle liquid outlet line was shut just before the tracer injection. In this test, it was also

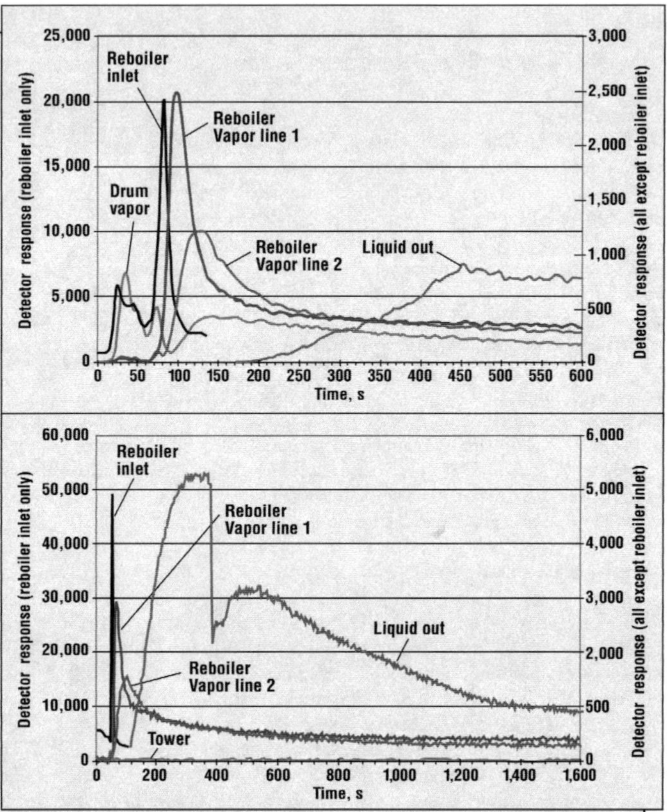

FIGURE 4. Detector response to a non-volatile tracer pulse test at 195 gal/min feed, conclusively shows heavy entrainment from the reboiler inlet area and provids means of measuring the entrainment, with (a; top) drum-drain valve open and (b; bottom) drum-drain valve closed for first 400 s

desired to positively verify that all the tracer remained in the liquid. To achieve this, the detector in the drum vapor outlet line (green) was moved to the depropanizer, and mounted in the vapor space below the bottom tray. Figure 4b shows no tracer on this detector, verifying that no tracer was in the vapor phase.

Figure 4b shows similar trends to Figure 4a in the first 200 s. Tracer peaked at the reboiler inlet 60 s after injection (black), followed immediately by a peak in Vapor line 1 (red). At about the same time, tracer was first detected in Vapor line 2 (blue), peaking 40 s later. Tracer was detected at the kettle liquid outlet after 120 s, peaking and then leveling off after about 300 s.

With the tracer in the reboiler outlet liquid steady at about 5,200 counts, the valve on the 3 in. drain-line from the drum was opened at 400 s. Immediately the detector count dropped, stabilizing at about 3,100. From that drop of counts, the flowrate of liquid from the disengagement drum can be calculated. A component balance on the tracer gives the following:

(Continues on p. 31)

ances equate the heads between the centerline (CL) of the 8-in. liquid line to the reboiler (lower reference elevation), and the centerline of the 10-in., reboiler-return nozzle (upper reference elevation).

The force on the tower side equals the liquid head measured in the depropanizer (above the lower reference) and the static head of the vapor between the liquid level and the upper reference. The force on the reboiler side is the total of three groups. The first sums the friction heads in the reboiler circuit. The second sums the liquid static head in the reboiler inlet line (from the lower reference to the reboiler entrance, the vapor static head in the vapor space between the reboiler liquid and the upper reference), and the mixed vapor-liquid static head in the vapor lines from the kettle to the drum. The third group is the reboiler liquid head (reboiler baffle height plus the liquid head over the weir), multiplied by an aeration factor. The head over the weir is calculated from the Francis weir formula [*10*]. Adding the three groups gives the total head from the reboiler side, which should equal the liquid head on the tower side.

The aeration factor of the kettle liquid is central to kettle force balance. Non-aeration is the conservative assumption often applied for design, as recommended by key literature references [*11*, *12*]. Comprehensive models [*13*, *14*] take into account the generated vapor volume and the bundle two-phase pressure drop. Recirculation of liquid inside the kettle and the existence of clear liquid zones between the

TABLE 2. REBOILER FORCE BALANCES						
Case	140 gal/min		195 gal/min		210 gal/min	
% Liquid in kettle	100	86	100	86	100	86
Entrainment removed in drum, lb/h	—		13,600		14,650	
Entrainment returned to tower, lb/h			17,600	30,800	95,400	106,200
Total liquid to kettle, lb/h	91,000		146,300	159,500	234,000	244,800
Vapor from kettle, lb/h	70,100		96,100	96,100	103,500	103,500
Wt. % of total liquid vaporized	77		66	60	44	42
Force balance, column side, ft liquid						
Level in tower above reference[1]	4.67		6.25		8.75	
Vapor static head above liquid[2]	0.52		0.34		0.06	
Total head, ft liquid	5.19		6.59		8.81	
Force balance, reboiler side, ft liquid						
Friction in inlet liquid line	0.27		0.69	0.83	1.78	1.94
Friction in vapor lines, kettle to drum	0.04		0.11	0.12	0.2	0.22
Friction in vapor line, drum to tower	0.27		0.7	0.79	1.45	1.54
1. Total friction	0.58		1.5	1.74	3.43	3.7
Static head in inlet liquid line	2.09		2.09	2.09	2.09	2.09
Vapor static head, drum and kettle	0.24		0.24	0.24	0.24	0.24
Mixed static head in vapor lines	0.32		0.51	0.57	0.79	0.83
2. Total static heads	2.65		2.84	2.9	3.12	3.16
Kettle head over weir	0.08		0.07		0.08	
3. Aerated liquid head	2.27	1.96	2.26	1.95	2.27	1.96
Total head [1+2+3], ft liquid	5.5	5.19	6.59	6.59	8.81	8.81

1. Reference: Centerline (CL) of liquid line to reboiler.
2. Head from liquid level to centerline of reboiler return nozzle.

bundle and the sides of a kettle reboiler complicate such models [*13*]. The force balance for the 140 gal/min test, in which the entrainment was small (as established earlier), allowed us to derive the aeration factor based on a good measurement. This is far superior to expert opinions and mathematical models.

For the 140 gal/min test, the liquid head calculated on the tower side was 5.19 ft (Table 2). Based on non-aerated reboiler liquid, the reboiler side liquid head is about 4 in. higher. The kettle liquid aeration factor was then varied to close the force balance. This calculation (Table 2) yielded an aeration factor of 0.86 (in other words, the reboiler material below the reboiler baffle was 86% liquid, 14% vapor).

A 4-in. difference out of 5.2 ft may appear insignificant, but of this 2.09 ft is liquid head above the lower reference (equal on both sides), and another 0.6 ft is static head differences, which have little error. So the basis for the aeration factor is 4 in. out of a net head of 2.5 ft. The friction head, which is the least certain term, is only 0.58 ft. So we expect the calculated aeration factor to be a good approximation, not a mere reflection of measurement or calculation errors.

For the 195 gal/min test (Figure 5), a trial and error calculation determined the quantity of entrainment (E; lb/h) from the drum to the tower bottom sump that would close the force balance. This is in addition to the 13,600 lb/h entrainment knocked out by the drum and directed to the reboiler liquid outlet, which was measured as described earlier. The entrainment rate E that reached the bottom sump (Table 2) was at least 17,600 lb/h (assuming non-aerated kettle liquid), and most likely was closer to 30,800 lb/h (using the 0.86 aeration factor determined above).

The last force balance in Table 2 is for the 210-gal/min depropanizer feedrate. No test was performed at this rate. However, plant experience has shown that whenever the depropanizer feedrate exceeded 205 gal/min, the liquid level rose above the reboiler return nozzle and initiated column flooding. The force balance basis was on the experience that at 210 gal/min the liquid level was at the bottom of the reboiler return nozzle.

The flows through the reboiler circuit for this calculation were prorated from the 195 gal/min feed, an increase of 7.7%. The entrainment rate that closed the force balance was about 100,000 lb/h, regardless of whether or not an aeration factor of 0.86 was used.

For the 195 gal/min test, the total friction head was 1.74 ft. Prorating for a 7.7% increase in flowrates, the friction would have risen by about 16%, just less than 0.3 ft. The observed increase in liquid head was 2.2 ft. The only conceivable explanation is a massive increase in entrainment. The force balance in Table 2 estimates a huge entrainment increase from about 31,000 lb/h to 106,000 lb/h over this small 7.7% increase in tower feedrate.

In summary, the force balances revealed that the kettle liquid was about 86% liquid, 16% vapor at the 140 gal/min test. The entrainment rate from the drum to the tower at the 195 gal/min test was about 31,000 lb/h, jumping to 106,000 lb/h once the tower feedrate was raised to 210 gal/min.

Heat transfer calculations

Table 3 shows the process data measured and calculated for the kettle. Most process temperatures were measured. There was no flowmeter, nor outlet temperature measurement, on the heating medium, so the reboiler duty (and therefore the heating medium flowrate and the heat transfer coefficient) were based on the depropanizer energy balance via the tower simulation. The entrainment rate from the drum is based on the rise in tower base level when the valve was shut, which we consider to be the most reliable of the measurement methods used.

The heat transfer coefficient could not be determined for the 140 gal/min test because the outlet temperature of the heating medium was not measured. In the 195 gal/min test it was measured by the pyrometer. The heat transfer coefficient measured for the test was 96 Btu/h-ft^2-°F as compared to 95 Btu/h-ft^2-°F calculated from the HTRI software. This perfect agreement is fortuitous, but also confirms good heat transfer performance.

The kettle boiling zones

Our measurements discussed above clearly identified two boiling zones in the kettle: a colder, intense boiling region that produces significant entrainment, stretching between the reboiler

TESTING METHODS (

Continued from p. 29)

$$3{,}100\ W = 400\ X + 5{,}200\ (W - X) \quad \textbf{(1)}$$

Where W is the flowrate of the C_4+ stream, known from the downstream debutanizer mass balance, and X is the flowrate of disengagement drum liquid, unknown. The count of 400 was obtained from the drum vapor inlet detectors once the bottom count leveled at 3,100 counts. This is not an accurate representation for the drum liquid, but being small, the calculation is insensitive to its accuracy. Solving Equation (1) gives $X = 62$ gal/min (hot), or 44% of the C_4+ stream.

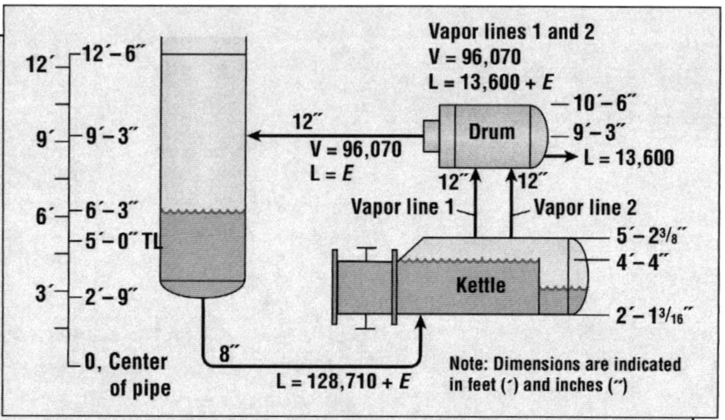

FIGURE 5. Reboiler force balances were applied to determine kettle aeration and entrainment rates (*E*; 195 gall/min test shown here; flows are in lb/h; TL = Tangent line)

This test produced an unexpected fallout. When the valve in the drum drain line was closed, the liquid level in the depropanizer tower started shooting up. Instead of being removed by the now-closed drum drain line, the liquid was carried over into the tower base. Over the seven minutes at which the valve was closed, the level glass indicated that about 55 ft^3 of liquid accumulated in the tower base, giving an average rate of 59 gal/min. This figure was strikingly close to that inferred from the tracer component balance above.

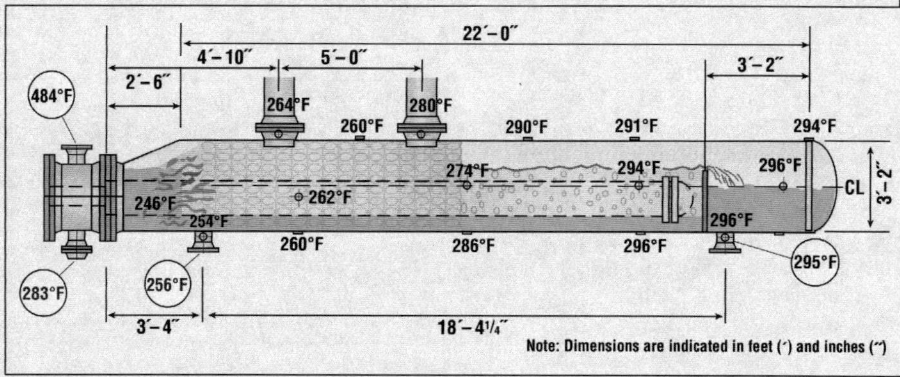

FIGURE 6. Analysis of boiling regions in the depropanizer kettle reboiler revealed that the inlet region was thermosiphoning, while the outlet region operated like a kettle

One more way was found for calculating the drum liquid flowrate. From the composition of the C_4+ stream (Table 1) and the reboiler liquid outlet temperature, the bubble point pressure was calculated. For the 140 gal/min test, the reboiler liquid outlet temperature was 270°F (Figure 2a), giving a calculated bubble point pressure of 275 psia, close to the measured reboiler pressure of 280 psia. The close agreement suggests that the C_4+ stream for this test consisted mostly of reboiler outlet liquid, and that the liquid stream from the drum is small.

For the 195 gal/min test, the reboiler liquid outlet temperature was 295°F (Figure 2b). The bubble point pressure at this temperature and the C4+ composition (Table 1) was 324 psia, well above the measured reboiler pressure of 280 psia. The C_4+ stream was therefore much lighter than the reboiler outlet liquid. This can only happen if a large portion of the C4+ stream came from the lighter drum liquid. A component balance gives

$$W x_w = X x_x + (W - X)\, x_k \quad \textbf{(2)}$$

Where W is the flowrate of the C_4+ stream, known from the downstream debutanizer mass balance; x_w is the component concentration in the C_4+ stream, known from the analysis; X is the flowrate of drum liquid, unknown; x_x is the component concentration in the drum liquid, known from the analysis; and x_k is the component concentration in the kettle outlet liquid, unknown. The composition of the kettle outlet liquid x_k at the reboiler liquid outlet temperature needs to give a bubble point pressure equal to the reboiler pressure. Equation (2) and the bubble point criterion were solved simultaneously, giving $X = 79$ gal/min, in reasonable agreement with the preceding alternative methods.

Composition and temperature behavior

Equation (2) sheds light on the movement observed in the kettle temperatures (by as much as 8°F near the outlet) in the 195 gal/min test. When the drum liquid rate X decreases in Equation (2), more lights reach the kettle outlet liquid, and its boiling point decreases. Therefore, the movement in kettle temperatures reflects variability in entrainment rate. This may also explain why the bubble point calculation gave a higher drum flowrate than the tracer and tower bottom level rise methods. In the 140 gal/min test, the entrainment rate was small, and the temperatures did not vary.

The above also explains the larger temperature difference between the kettle outlet and inlet liquid temperatures in the 195 gal/min test. Much of the light kettle liquid was entrained, making the remaining liquid much heavier, thereby increasing the temperature difference across the kettle.

The fact that 44% of the C_4+ was drum liquid explains why the drum liquid approached the C_4+ composition in the 195 gal/min test. Recycle of a large amount of entrainment, much of it not captured by the drum, accumulated heavies in the reboiler circuit. Also, more entrainment came from the region of the reboiler between Vapor line 2 and the liquid draw, as can be inferred from the larger difference between the temperatures of Vapor lines 2 and 1 in the 195 gal/min test. ❑

liquid-feed nozzle and Vapor line 2, and a low entrainment, hotter region between Vapor line 2 and the kettle liquid-draw nozzle. These are very similar to two of the three regions identified by Rubbers et al. [3] in an extractive-distillation kettle reboiler. The only region identified by Rubbers that was not clearly observed in this investigation is the stagnant liquid zone near the inlet tubesheet. The discussion below considers the application of the zones identified by Rubbers to those identified in the depropanizer reboiler (Figure 6):

Quiescent liquid region. Rubbers identified a stagnant region or dead zone in the tubesheet quadrant. With little disengagement space, not much boiling takes place, light components flash off, and the liquid becomes rich in the non-volatile components. The non-volatile liquid with little aeration incurs a high hydrostatic head, which resists movement of fresh liquid into this region. The reboiler surface temperatures measured by Rubbers in this region were high and close to the heating medium, which supports stagnation.

The average aeration of 86% liquid, determined from the reboiler force balance for the 140 gal/min test, is close to non-aerated liquid, but not quite there. So it appears that there is a relatively stagnant, or so-called quiescent-liquid zone in the depropanizer reboiler, but it was not completely stagnant.

It is important to recognize that in the extractive distillation case of Rubbers, the reboiler liquid consisted largely of a high-boiling solvent, so once the light components were flashed, the reboiler ΔT was quite small. In the depropanizer, the ΔT near the tube sheet was much larger, especially for some tube passes, so the two cases are not directly comparable.

The high L/V intense-boiling region. Rubbers observed a second region with higher-than-average upward liquid traffic to compensate for reduced flow to the stagnant region. This resulted in colder liquid temperatures but intense boiling and higher-than-average vapor generation. In the depropanizer reboiler, we too observed this region between the reboiler feed inlet and Vapor line 2.

A pressure balance across the reboiler equates the pressure difference between the bottom of the kettle and the reboiler's common vapor space, no matter which region one travels through. Along each vertical line from the reboiler bottom to the vapor space, the pressure difference is the sum of the liquid head plus the friction head. On the average, the pressure difference approaches the aerated liquid head. In the quiescent regions, the friction pressure drop is small due to the low flow. Conversely, in the intense boiling region, the intense boiling highly aerates the liquid and reduces the froth density (as evidenced from the neutron scan in Figure 3), which lowers the liquid head. This induces a high liquid flow into the region until the friction terms balance the difference between the less-aerated liquid head in the quiescent region and the highly-aerated liquid head in this region.

The vigorous boiling concentrated in this region means that the fraction of the liquid vaporized is lower, so less lights are removed from the liquid. This lights-rich liquid persists all the way to the two vapor nozzles, as evidenced from the neutron scans and the tracer tests. The lower temperatures observed in this region (Figures 2a and 2b) are the result of the high liquid flux in the tube field. The vapor nozzle closest to the tubesheet is coldest, because that is where the intense boiling starts, is more vigorous, and therefore leads to a smaller fraction being vaporized (also observed by Rubbers). At the same time, some of the liquid is projected into the vapor outlet nozzles as entrainment. The action in this region has similarities to the action in horizontal thermosiphon reboilers in that the quantity of vapor generated results in a low density froth, which promotes liquid flow to this region as opposed to the denser liquid in the less aerated zones.

The kettle boiling region. Rubbers observed a third region that actually operated as a kettle is intended to operate. In the depropanizer reboiler, we too observed this region between Vapor line 2 and the kettle liquid draw.

There is a net flow of liquid from the reboiler feed region towards the overflow baffle since this is the only manner (other than entrainment) in which the unboiled liquid can permanently exit the kettle reboiler. As the liquid moves toward the overflow baffle, it progressively contains less lights. This results in less vigorous boiling and a denser froth in this region. This was actually observed in the neutron backscatter (Figure 3). This denser froth provides more resistance to liquid from the kettle bottom and, in turn, reduces the liquid flow to the kettle boiling region.

The kettle boiling region is the one region that actually operates like it should — that is, as a kettle. There is not much entrainment here, as verified by the neutron backscatter (Figure 3), and there is a lot of disengagement space in this region, both of which promote true kettle behavior. As the flow of reboiler feed liquid through this region is relatively low, and the lights gradually disengage along the path from Vapor line 2 to the overflow baffle, the temperatures in this region gradually increase as observed by the temperature surveys (Figures 2a and 2b).

Can kettles thermosiphon?

The 195 gal/min test established conclusively that there was a large amount of entrainment from the kettle. While the entrainment knocked out in the drum was measured at 13,600 lb/h, about 30,800 lb/h entrainment continued from the drum into the tower bottom. In total, the entrainment rate for this test was 46 wt.% of the amount vaporized in the reboiler.

Proprietary software supplied by technology providers often includes an entrainment calculation for kettle reboilers. A key parameter used in the software we applied is the foam height above the overflow baffle. According to that software, to get a ratio of 0.46 by weight of the entrained vapor to amount vaporized, the foam height would need to be 3 in. below the outlet nozzles. For the 210-gal/min feed experience, the force balance (Table 2) gives a ratio of 1.03 lb entrained per lb vaporized. The entrainment calculation shows that for this ratio, the foam height needed to increase only 1–2 in. below the vapor nozzle.

Once the distance between the top of the foam and the outlet nozzle reaches 2–3 in., a kettle does not operate as a *bona-fide* kettle reboiler. It operates as a thermosiphon reboiler.

Table 2 shows the percent vaporization of the reboiler inlet liquid, as calculated from the force balances. For the 195 gal/min test, the vaporization rate was 60% by weight, declining to 42% at 210 gal/min. The latter, and even the former, vaporization rates are within the range of thermosiphon reboilers operation. So, as the entrainment escalates, so does the thermosiphoning action. The interaction between the level in the tower and the liquid circulation in the reboiler is also typical of thermosiphon reboilers, as described in Ref. *15*.

There is one large difference between the depropanizer kettle and a thermosiphon. In a thermosiphon reboiler, the liquid level in the tower is controlled, either by the draw rate (directly or indirectly), or by a constant head baffle in the bottom sump. This is not the case

TABLE 3. REBOILER PROCESS DATA

	140 gal/min test	195 gal/min test		140 gal/min test	195 gal/min test
Heat duty, MM Btu/h	7.79 [3]	10.74 [3]	Drum vapor out, °F	257 [1]	272 [1]
C_4+ flowrate, lb/h	21,060 [1]	32,640 [1]	Reboiler process liquid out, °F	270 [1]	295 [1]
Amount vaporized, lb/h	70,050 [3]	96,070 [3]	Heating medium in, °F	489 [1]	484 [1]
Reboiler feed [excluding entrainment], lb/h	91,110 [3]	128,710 [3]	Heating medium out, °F	ND	283 [2]
			Heating medium flow, lb/h	ND	81,000 [5]
Entrainment knocked out in drum, lb/h	ND	13,600 [2]	Reboiler pressure, psia	280 [1]	280 [1]
			LMTD, °F	ND	83 [4]
Entrainment from drum, lb/h	ND	30,800 [6]	Heat transfer coefficient, MM Btu/h-ft^2°F	ND	96 [5]
Reboiler process in, °F	245 [1]	256 [1]			

Notes:
1. Measured by process instrumentation.
2. Measured by pyromenter or field technique.
3. Calculated from simulation.
4. Calculated from process measurements.
5. Calculated from heat duty and LMTD.
6. Calculated from reboiler force balance.
ND. Not determined.

in a kettle reboiler, where the product draw is controlled by the overflow in the kettle. This overflow has no relation to the tower base level. As the liquid head in the tower rises, the percent vaporized declines, so the liquid fraction carried over rises. As it rises, the pressure drop rises rapidly, which further increases the level, which in turn further increases the entrainment, and so on. Without a level control this response is unstable and can run away.

The force balances provide evidence that the runaway mechanism above occurs in the depropanizer kettle. The force balances show that without a sharp escalation in entrainment, the friction pressure drop should have raised the depropanizer bottom level by only 0.3 ft when going from 195 gal/min feed to 210 gal/min feed. The observed increase of that level, however, was more than 2.2 ft.

The extent to which the runaway mechanism restricts the reboiler circuit capacity can be tested by drawing product from the bottom of the tower. This test is currently under consideration. If successful, this idea is likely to permit the reboiler to operate as a thermosiphon at higher throughput. While this may not be ideal, there is no fundamental reason why a kettle reboiler should not be able to work as a thermosiphon once the runaway side effect is eliminated.

So to answer our question: yes, indeed, kettle reboilers can turn into thermosiphons.

Lessons learned

The root cause of the kettle turning into a thermosiphon is entrainment. Without mega-entrainment, there is not enough liquid to raise the liquid level in the tower base to the reboiler return nozzle elevation.

The key to preventing entrainment is to follow good design practices. The criteria recommended in the literature (and described earlier in the introductory section) should be followed. In the depropanizer reboiler, many of these criteria were violated. Specifically, the height between the overflow baffle and the top of the shell was 10 $^5/_{16}$ in., which is far too short. The solution recommended for the depropanizer is a larger reboiler, in which the vapor space above the overflow baffle will be large enough to prevent excessive entrainment.

The disengagement drum performed very well. It was found to remove about 60 gal/min of entrainment. Without it, the reboiler bottleneck would have occurred earlier. Adding a disengagement drum above an entraining kettle reboiler can be a good "band-aid" fix that will obtain more capacity.

Last, but not least, one good measurement is worth 1,000 expert opinions. The more good measurements, the better the troubleshooting diagnosis. Using alternative measurement techniques affords invaluable cross checks and validations. The winner is the conclusive diagnosis and a good path forward towards a solution. ■

Edited by Rebekkah Marshall

Acknowledgement

The authors wish to express their gratitude to Tracerco, and Mike Boone in particular, for superb gamma and neutron scans and tracer tests, which were invaluable in our endeavors. Mike also contributed many testing ideas that helped obtain invaluable insight into the problem.

Authors

Henry Z. Kister is a senior fellow and director of fractionation technology at Fluor Corp. (Phone: 1-949-349-4679; Email: henry.kister@fluor.com). He has over 30 years experience in design, troubleshooting, revamping, field consulting, control and startup of fractionation processes and equipment. He is the author of three books, the distillation equipment chapter in Perry's Handbook, and over 80 articles, and has taught the IChemE-sponsored "Practical Distillation Technology" course more than to 350 times. A recipient of several awards, including this magazine's 2002 Personal Achievement Award, Kister obtained his B.E. and M.E. degrees from the University of NSW in Australia. He is a Fellow of IChemE and AIChE, and serves on the FRI Technical Advisory and Design Practices Committees.

Milton A. Chaves is a process engineer at Hess Corp.'s Tioga Gas Plant (Phone: 701-641-0096; Email: mchaves@hess.com). He has two years experience in design, troubleshooting and process optimization. Prior to his current position he worked for ABB and at the Hovensa refinery in St. Croix, U.S.V.I. A first-place award recipient of Texas A&M University's (TAMU) Spring Senior Plant-Design Competition, Chaves obtained his B.S.Ch.E. degree from TAMU in May 2008.

References

1. Kister, H. Z., "What Caused Tower Malfunctions in the Last 50 Years", Trans. IChemE, Vol. 81, Part A, p. 5, January 2003.
2. Kister, H. Z. "Distillation Troubleshooting", John Wiley and Sons, NJ, 2006.
3. Rubbers, E., Green, K., Fowler, T., Kister, H. Z. and Stupin, W. J., "Distillation Reboiler Startup Can Pose Challenges", *Chem. Eng.*, p. 55, February 2004.
4. Kern, D. Q. "Process Heat Transfer", McGraw-Hill, NY, 1950.
5. Palen, J. W. and Small W. M., A New Way to Design Kettle and Internal Reboilers, *Hydroc. Proc.* 43 (11), p. 199, 1984.
6. Shah, G. C., "Troubleshooting Distillation Columns", *Chem. Eng.*, July 31, p. 70, 1978.
7. Lord, R. C., Minton, P. E. and Slausser, R. P., Design Parameters for Condensers and Reboilers, *Chem. Eng.*, March 23, p. 127, 1970.
8. Tammami, B. Simplifying Reboiler Entrainment Calculations, *Oil and Gas J.*, July 15, p. 134, 1985.
9. Hartman, E. L., and D. W. Hanson, Reboilers Behind Distillation Tower Problems, p. 111, *PTQ*, Spring 2001.
10. Green, D. W., and Perry, R. H. "Perry's Chemical Engineers' Handbook" 8th Ed., Section 14, McGraw Hill, New York, 2008.
11. Lieberman, N. P., "Process Design for Reliable Operation", 2nd Ed., Gulf Publishing, Houston, 1988.
12. Palen, J. W. and Small, W. M., Kettle and Internal Reboilers, *Hydroc. Proc.*, 43 (11), p. 199, 1964.
13. Fair, J. R. and Klip, A., Thermal Design of Horizontal Reboilers, *Chem. Eng. Prog.*, p. 86, March 1983.
14. Ishihara, K., Palen, J. W. and Taborek, J. Critical Review of Correlations for Predicting Two-Phase Flow Pressure Drop across Tube Banks, *Heat Trans. Eng.*, 1 (3), p. 23, January–March 1980.
15. Kister, H. Z., "Distillation Operation", McGraw-Hill, New York, 1990.

YPF

Fractionation towers Inspection from the point of view of the process

Introduction

This paper was written to share experiences about fractionation towers inspection from the point of view of the process. The items mentioned are intended to provide guidance to the future towers inspectors to focus on points extremely important and relevant that will impact directly over the operation of distillation columns.

Inspection of the column components is usually made by the Equipment Inspection Department. The Production Department, besides checking if it is necessary to replace or repair internals, inspects if the tower is clean or it has accumulated large quantities of coke, salt, or corrosion products during operation. Cleaning, repairs and/or mountings are the responsibility of the Maintenance Department.

The Process Department inspector is responsible for critically analyzes the changes made to make in the column and for identifying future operational problems. Something as simple as the replacement of a downcomer can generate a significant fractionation efficiency loss if is it not properly mounted. For example, if the clearance under downcomer is too small, the column could suffer from flooding, because of the flow restriction. The construction crew do not always understand how the column works. And therefore may often make decisions which are not the best, precisely because of their lack of awareness of how a tower works. It is essential to have people that analyze critically and in detail the column to address this point. In addition to the critical analysis, it is imperative that the process inspector avoid copying usual old practices and errors that can be dragged for years, becoming part of the way of working.

A critical inspection is a tedious and often dirty task. It is necessary to remain inside the tower for hours. However, it is a great opportunity to identify potential problems that otherwise would affect the column when returning to service. If we make a good job, we will save headaches when the tower is already in operation.

It is important to identify problems before opening the fractionation tower for inspection. The monitoring of the variables of the column will allow focusing on sectors that have problems in advance. For example, if the column is flooded at the top, between the kerosene sump and the top of the tower, (Identified by the differential pressure and gamma scanning), when opening the column, immediately you should plan to inspect that section. In our particular case, it was intended to replace the distributors and the wash bed of the vacuum tower, because we knew from monitoring pressure drop during operation that it was filled with large quantity of coke.

Fractionation towers Inspection from the point of view of the process covers the previous examples, and more. To put into practice this inspection it is necessary to prepare a detailed program to know what to inspect and how look for the potential problems.

An old Process Engineer told me once that good observer and detector of potential problems turns inside the column into liquid, that flows through the downcomers and into vapor, that rises by trays.

Background

This paper focuses on inspections performed in Vacuum and Atmospheric fractionation towers in our refinery, during routine maintenance shutdown of the units, on November 2008. The Refinery, called Luján de Cuyo (Belonging to YPF), is located in Argentina, in the province of Mendoza.

Our vacuum tower, CV10, is 37 meters length and a diameter of 8.2 meters in its largest section. It processes around 9000 m3/d of atmospheric residue. Our Crude Distillation

Atmospheric tower, CV1, is 39.6 m length and 4 m in diameter. It processes around 11000 m^3/d of crude.

Full inspections for each of the towers had been previously planned. Previous planning is necessary to avoid delays and time extensions of the scheduled shutdown. Although a complete inspection of each one of the towers was conducted, this paper presents only the points of major interest, which will be useful for process inspectors.

We will focus on the following points:

Vacuum tower:

- Inspection and testing of distributors.
- Inspection and testing of sumps and chimney trays.
- Inspection of dual-element filter screens.
- Design problems.

Atmospheric distillation Tower:

- Inspection of trays. (valves, holes, measurements of weirs and clearances under the downcomers).
- Inspection and testing of sumps.
- Design problems.
- Test of Gravity flow distributor.

VACUMM DISTILLATION TOWER

Light Vacuum Gas Oil (LVGO) pump around distributor

Among a total of 19 spray nozzles, only one was found plugged.
It was agreed with the maintenance staff dismounting 20% of all nozzles, selected at random to check. If 50% of these were plugged or dirty, all nozzles should be dismounted for cleaning and inspection.

The nozzle that was found plugged, was plugged with a piece of packing (Flexiring). To shed light on our question of how it could have gotten there, we went to inspect the dual-element filter screens, external to the tower.

We see two different hypotheses:

- We found no meshes inside the filters, allowing the packing piece to bypass them.

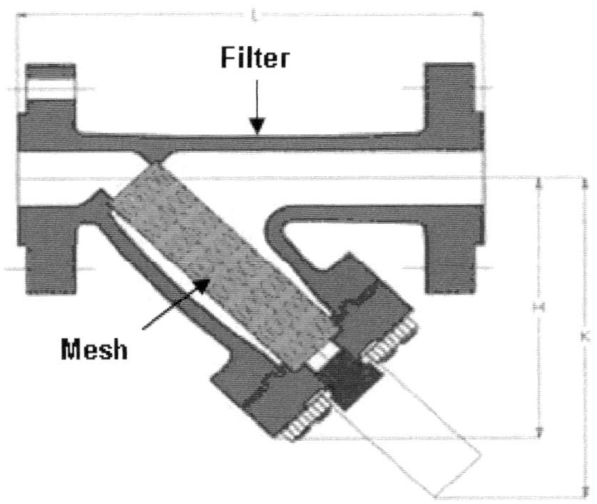

- One of the dual-element filters was dismounted to connect a segment of piping that bypassed the dual filters. Somebody decided to dismount it to give operational flexibility to the unit. The next drawing shows the function of this pipe.

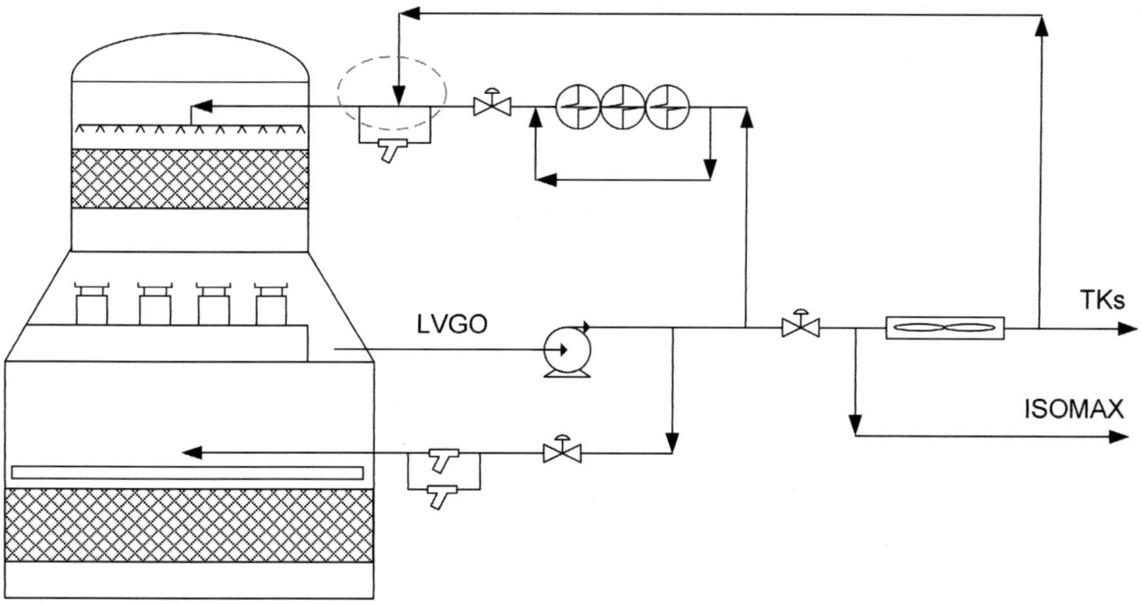

When exchangers are shut down for cleaning or repairing because of an internal damaged of one of them, It is necessary to bypass all of them. It is not possible to bypass only one. If this happens, the unit must be shut down. Connecting a pipe from the air coolers gave more operational flexibility, but is not a good practice for to the distributors protection.

Distributor testing

To verify that no spray nozzles have been plugged, they were tested with water.

In the following pictures one can see clearly what happens when the flow is not adequate, mainly due to operational failures, or in the worst case, plugging of the nozzles.

A plugged nozzle produces a poor flow for washing on a portion of the bed.

LVGO / Heavy Vacuum Gas Oil (HVGO) fractionation bed distributor

We have a gravitational distributor in the LVGO / HVGO fractionation bed.
The LVGO fills the pre-distributor channel through feed pipe. The liquid overflows into parting BOX. Then, troughs are filled. At the top of each trough there are orifices, where the liquid flows to drip tubes. The liquid arrives to spoons base and distributes liquid to the bed.

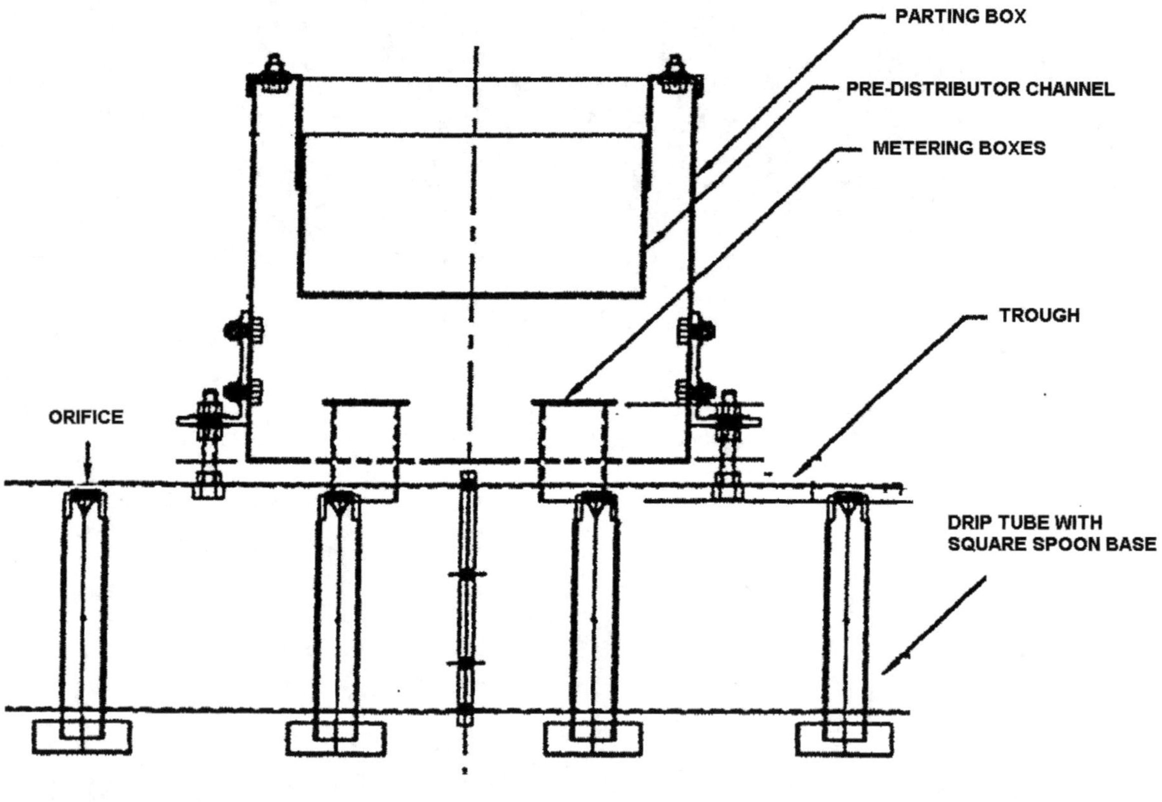

The first thing that was observed during the inspection of this distributor was that the central parting box was uneven. The next photograph shows a missing bolt necessary to holding the box.

Unevenness in the central parting box will affect liquid distribution to the bed below, since the liquid will flow preferentially to some troughs. A section of the packing will be dry due to reduced irrigation, and poor end point or cloud point would be an issue.

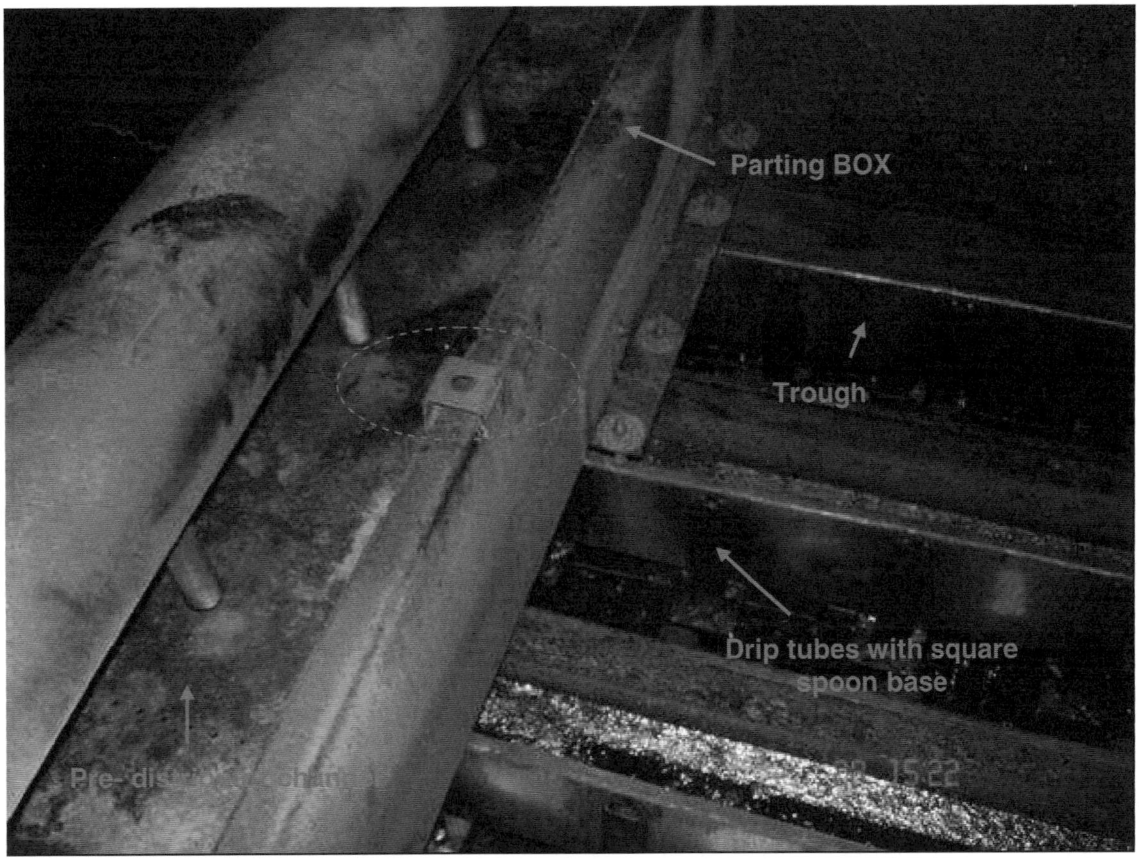

In certain sections, drip tubes with square spoons base were not in place. They were on the bed. Also, as shown in the picture, some were bent so that the liquid distribution to the bed was not good.

Even though to enter the column it was (and should be) washed for hours, there was dirt inside the side troughs of the distributor. As the holes from where the liquid issues are above where the dirt was in the troughs, there should have been no obstruction to the passage of liquid. This shows the effectiveness of this design in terms of antifouling.

While one of the advantages of using this technology is due to the little dirt, we should not trust at all when inspected. The photograph below shows a clear blockage in the center parting box, which hindered the passage of liquid to the lateral troughs. This was discovered through a water test of the distributor. As a result, the distributor was dismounted and cleaned.

Heavy Vacumm Gas Oil (HVGO) pump around

We found a design mistake. The pictures below show how the branch pipe was wetted by liquid issuing from the spray nozzle. This occurred because the pipe was inclined.

Given that the values of total acid number (TAN) and sulfur content of the current HVGO are 1.3 kg [mgKOH / g HC] and 0.67%w/w respectively, and the operating temperature exceeds 200 °C, the jet flow from the nozzle could negatively impact and contribute to corrosion of the feed piping.

We placed a longer nozzle to avoid the liquid wetting the branch pipe. The tube placed was as long as we could make.It was impossible to install a longer one due to the inclination of the pipe. Pictures below show how the spray distribution would have wet the branch pipe weld if we decided to carry out the modification.

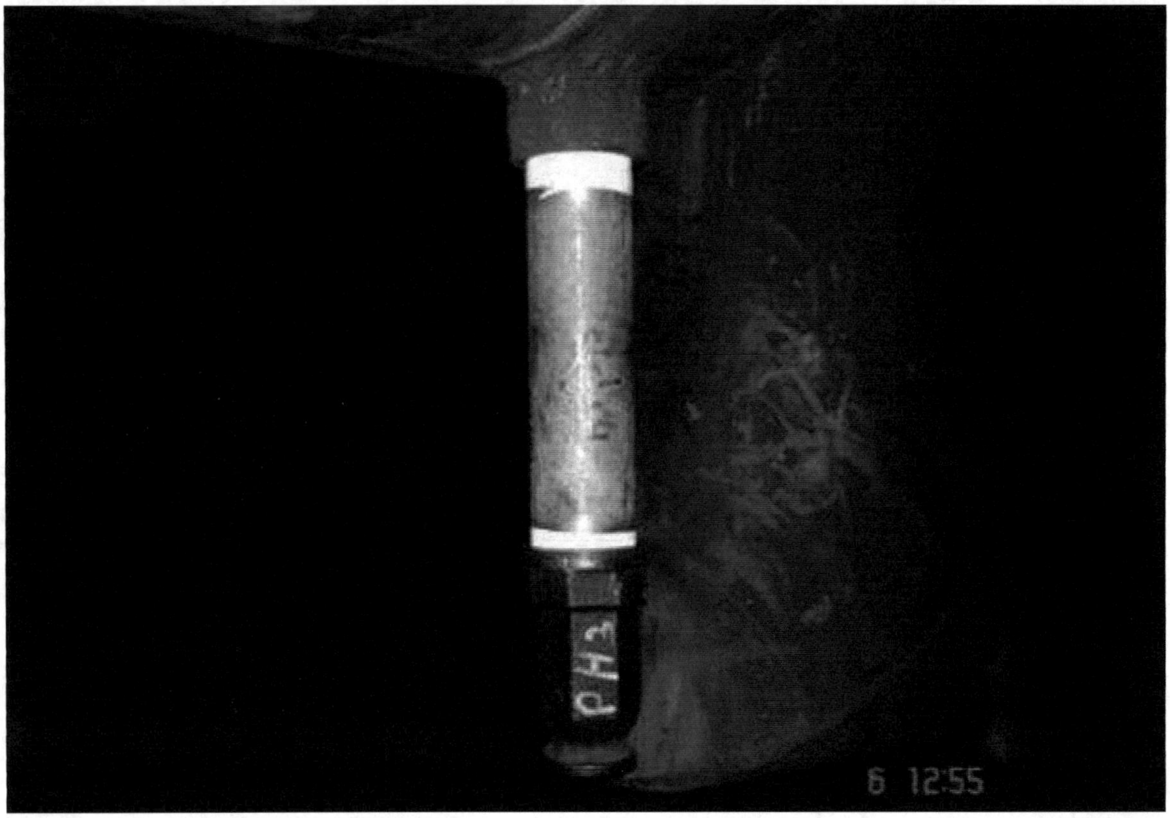

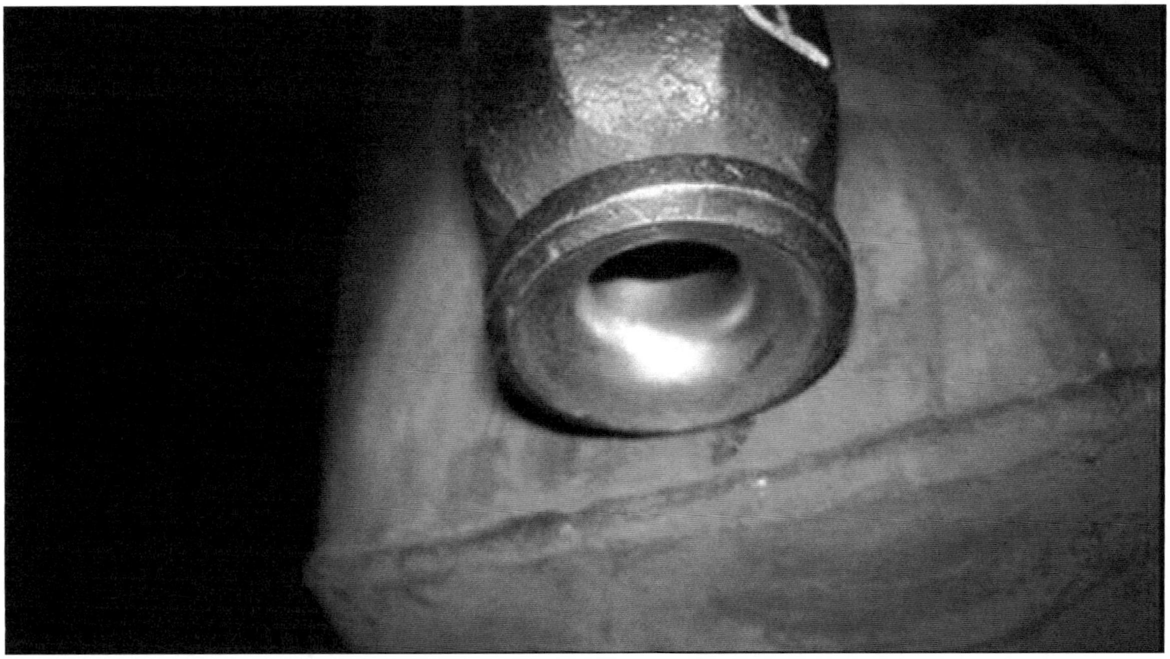

Once we saw this, we became concerned about corroding the weld. We therefore decided to go back to the original arrangement. Any leak in branch pipe would cause maldistribution over the entire bed, leading to poor heat transfer and possibly coking on the packing. Our decision therefore was to mount the shorter former pipe with the compromise to replace it during the next shutdown of the unit.

While no nozzles were plugged, it is appropriate to comment on another design error as to the dual filters of the HVGO pumparound. They are located upstream of the control valve. Dual filter screen should be downstream of any equipment before entering the column, to avoid plugging in the nozzles. This anomaly dates from the construction of the plant in 1970. We will fix them during the next shutdown of the unit.

Wash bed distributor

Before entering the tower for inspection, we observed from the manway an arm of the distributor that was broken, so liquid came out between the two sections.

The visual inspection, inside the column, showed 8 of 61 plugged nozzles. In this particular case, all the spray nozzles were replaced by new ones, so all of them were dismounted. When we inspected the dismounted nozzles, we were surprised to find that more than 8 nozzles were plugged. About 50% of the nozzles were plugged internally. This would have probably appeared during the water test of the spray nozzles. The lesson in this case is the importance of the spray distributor water test.

Regarding filters "Y", we found only 1 whose mesh was deformed, allowing the passage of dirt to the spray nozzles.

Filter.

Mesh.

The next photograph shows how the upper grid of the bed was raised to the height of manway. Close to the manway we found a broken tie rod.

The following picture shows how grids were deformed.

This bed was entirely replaced as planned.

One thing to consider is that when the maintenance staff mounted this new bed, they were doing without drawings. According to design drawings, they should have each layer at 45 ° to the layer below, and this was not being respected. So we had to intervene to properly reinstall the layers. The lesson is to try to keep abreast of all changes made. The photograph below shows a correct flow grid layer arrangment.

YPF

ATMOSPHERIC DISTILLATION TOWER

This column is composed by 39 valve trays, and a structured packing.

From top to bottom, the configuration is the following:

Trays #1 to #13

Tray #14 (Kerosene sump)

Trays #15 to #23

Tray #24 (Atmospheric Light Gas Oil sump)

Trays #25 to #33

Tray 34 (Atmospheric Heavy Gas Oil sump)

Structured Packing

Bottom trays #35 to #39. (Stripping section between petroleum feed and stripping steam inlet).

Trays from 1 to 10, are two-pass, and from 11 to 39, they are four-passes trays.

The inspection was conducted in detail for each of the trays and structured packing. This paper only shows the highlights.

Regarding the trays inspection, we found no valves in some cases along the column. Some valves were found corroded.

Improper assembly of the panels that compouse the trays, produce holes that allow weeping worsen the quality of fractionation.

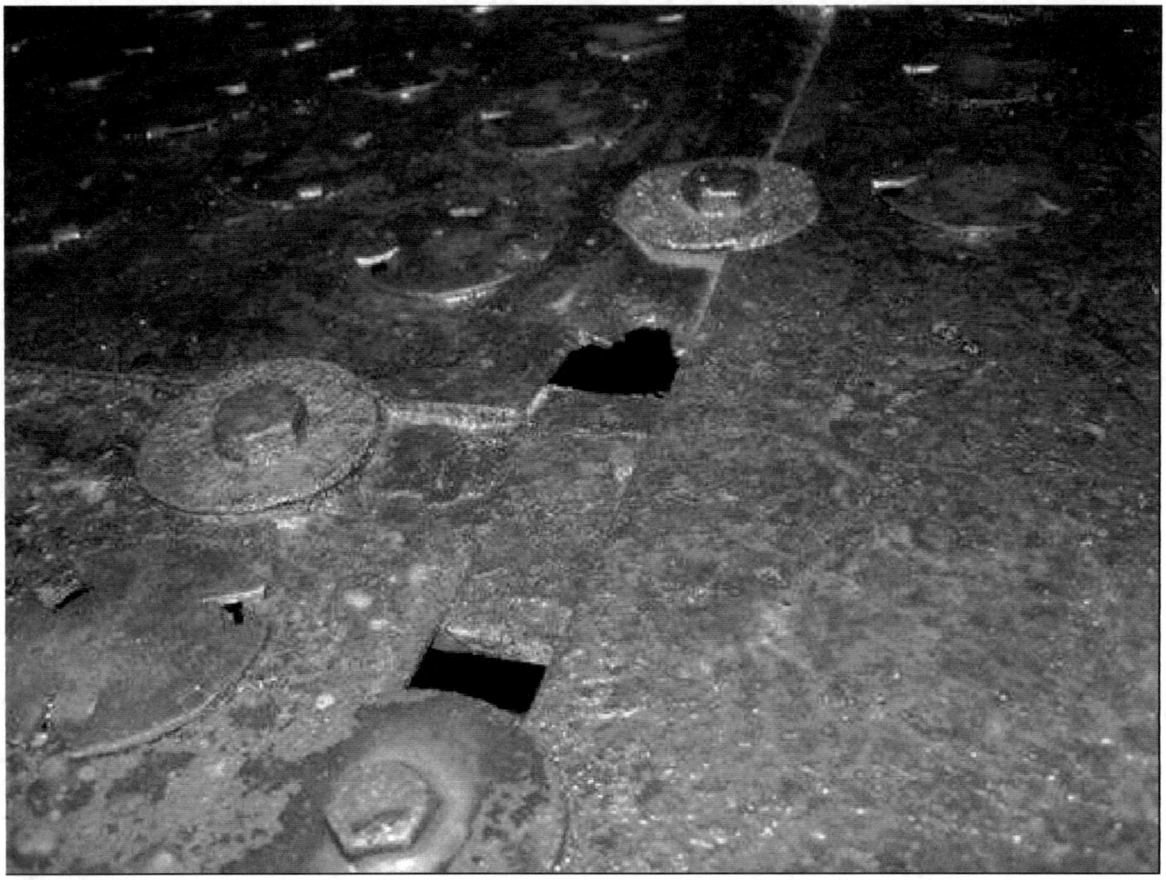

Besides the formation of holes, improperly placing a panel may lose the functionality for which it was built.

The next photograph shows a panel rotated about its normal position. As can be seen, if the panel is rotated, the inlet weir will not work. The inlet weir function is to prevent the liquid from the downcomer from lifting the valves closest to the inlet downcomer and weeping through the inlet row of valves.

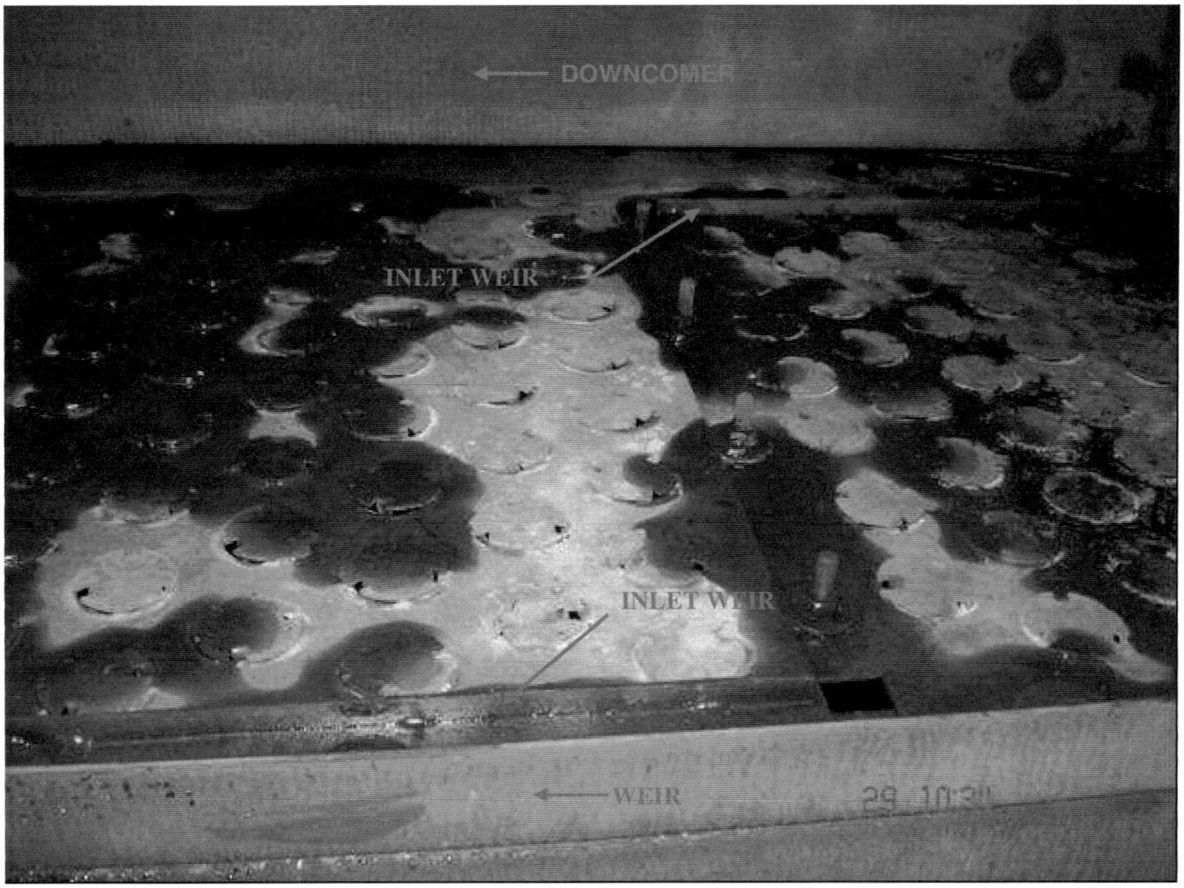

In each of the trays we took measurements of the weirs, and the clearances under the downcomers to ascertain a liquid seal on the tray. In case of not having the seal liquid, gas may rise up the downcomers, producing liquid entrainment and flooding. Conversely, if the clearance under the downcomer is too small, it will increase the height of liquid in the downcomer. When the downcomer backup reaches the tray above, flooding will occur. It is sufficient to have one undersized clearance to bottleneck an entire fractionator.

In the particular case below, the clearance under the downcomer was 29 mm. According to the design drawing, this clearance should be 43.5 + / - 3mm.

This problem was solved by cutting the bottom of the downcomer to acceptable values.

Between trays #19 and #20 is the Atmospheric Light Gas Oil (ALGO) pumparound inlet pipe distributors. It was noted that while installing the distributors in the column, downcomers were damaged. This caused significant losses of fractionation capacity.

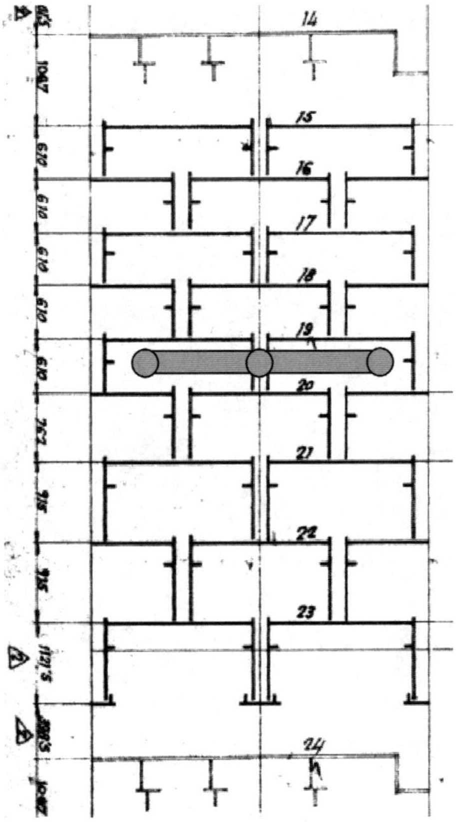

Side view of ALGO pump around inlet to the column.

Sumps

Kerosene, Atmospheric Light Gas Oil and Atmospheric Heavy Gas Oil sumps were tested with water. They were filled with water to a certain level. Then we cut off the water and waited about 20 minutes to look at the fall of the water level. The conclusion of the test was that no problems were observed with sump leakage.

Anecdotally, during the final visual inspection of the tower, before being closed, we found a pipe scaffold in the AHGO sump. It was hiding, so that if we had not properly performed the final inspection, the scaffold debris may have generated significant damage in the area.

Gravity flow distributor.

The gravity flow distributor, located in the wash section was tested with water, to see what happen when the flow increases. The next picture shows how it fills the central parting box.

Conclusion

It is very important that whenever a fractionation tower is open to inspect, it should be analyzed and inspected by a qualified process engineer who has been trained in what to look for.

Online Dynamic Parameter Estimation, Data Reconciliation and Transition Planning for a Packed Distillation Column within a Model-Centric Process Support System

Daira Aragón[1], José A. Romagnoli[2†]

[1]Dept. de Ingeniería Química, Facultad de Ingeniería, Universidad de Antioquia, Medellín, Colombia
[2]Dept. Chemical Engineering, Louisiana State University, Baton Rouge, LA 70803

The capabilities a framework for model-centric framework for support of process systems, integrating and easing the definition of offline simulation, parameter estimation, data reconciliation, and optimization problems has been extended to formulate and solve online activities related to parameter estimation, data reconciliation and transition planning. Different complex online dynamic parameter estimation, dynamic data reconciliation and transition planning problems were successfully solved when the framework was coupled to an OPC server for a packed distillation unit. This demonstrated the excellent capabilities of the framework for more advanced real-time applications.

1. Introduction

Today's world is experiencing events that demand changes in the goals and methodologies of research for the development and production of goods and commodities. The current globalization and tightening of environmental and safety laws, for example, require industries to have efficient production processes at a cost sufficiently low to stay afloat in this competitive market. Therefore, supporting tools and techniques are required so that process engineers are able to fulfill these demands. Computer-aided process engineering (CAPE) is the application of a modeling approach to the study of a process as an integrated whole (Mayer and Schoenmakers, 1998). Its objective is not only to promote the development of modeling and simulation tools, but also to assist in the integration of the process with its operating systems during design, as well as improve the mechanisms for data transfer between phases of the process life cycle.

Even though the central role of process models in cutting-edge technologies for advanced manufacturing operations cannot be ignored, rigorous mechanistic process models are just one of the many components of any sophisticated software tool targeting industrial applications. In order to bring the advances in modeling, simulation and optimization environments and open-software architectures closer to process industries, a series of novel mechanisms and advanced software tools must be devised so that the definition of complex model-based problems is simplified. Additionally, synergistic interactions between complementary model-based software tools must be refined in order to fully unlock the potential of model-centric technologies in the industrial workplace.

In previous works (Aragon et. al, 2008; Rolandi y Romagnoli, 2010) we have presented a framework for model-centric framework for support of process systems, integrating and easing the definition of offline simulation, parameter estimation, data reconciliation, and optimization problems. In this work, we extend the capabilities of the framework to formulate and solve online activities related to parameter estimation, data reconciliation and transition planning.

During process operation assisted by model-based applications, online parameter estimation and data reconciliation are indispensable because they provide the means to maintain an accurate model and to adjust measurements in accordance to such model. For online activities, such as advanced process control and real-time

† Correspondence concerning this article should be addressed to Jose A. Romagnoli. e-mail: jose@lsu.edu

optimization, maintaining a model that represents the process behavior accurately and reducing the magnitude of errors is essential as new trajectories are computed based on information provided by both process model and plant measurements. Additionally, transition planning (i.e. a special case of dynamic optimization) is extremely important for processes which require changes either in product specification or productivity.

A packed distillation unit has been selected to evaluate the performance of such online activities within the proposed framework. Modeling of tray and packed distillation columns is a common practice today. The use of fundamental modeling in model-based methodologies to packed distillation processes has been made possible due to advances in numerical methods and software applications. Temperature and flow measurements collected through the distributed control system (DCS) during the normal operation of a distillation unit at plant scale may serve for many different purposes such as process control, parameter estimation, data reconciliation and optimization, among others. However, the majority of studies in the area show the use of such information for control and optimization purposes (Bezzo et. al, 2005; Espinoza & Marchetti, 2007). Few authors report data reconciliation in distillation columns (Al-Arfaj, 2006; Bhat & Saraf, 2004) and even less report online dynamic parameter estimation and data reconciliation for these processes (Bai et al., 2007).

The framework for integrated model-centric process support formulated and solved different complex online dynamic parameter estimation, dynamic data reconciliation and transition planning problems when coupled to an OPC server for a packed distillation unit. This demonstrated the excellent capabilities of the framework for more advanced real-time applications.

2. The model-centric process support system

2.1 *Architecture and mechanisms*

The Integrated Model-Centric Process Support System (IMCPSS) responds to the need for software tools centered in the end-users (process engineers) for which the complexity of using current modeling environments is a barrier for broader application of model-based activities.

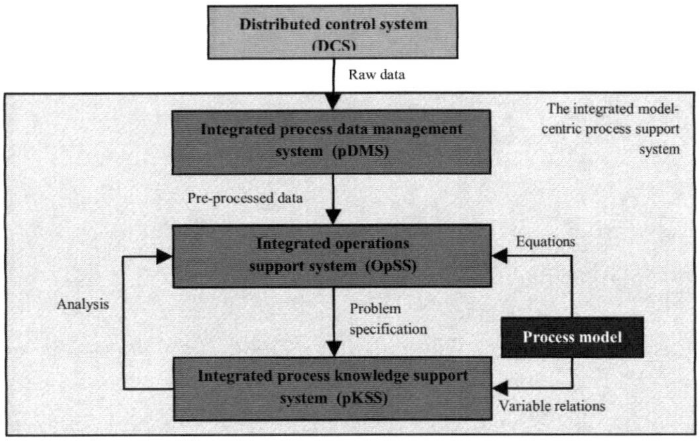

Figure 1. IMCPSS architecture

The proposed IMCPSS architecture (Figure 1) comprises three subsystems with particular purposes but in constant interaction: the *process data management system* (pDMS), the *operation support system* (OpSS), and the *process knowledge support system* (pKSS). The pDMS manages acquisition of plant data and prepares it for use within the OpSS. The latter supervises the problem formulation and execution of all included model-based activities. On the other hand, the pKSS validates that a problem is well-posed in terms of observability and redundancy. The interaction between subsystems of the IMCPSS has been achieved with the introduction of a novel component: the *Problem Definition Environment* (PDE). Practically, the PDE is a software component that combines tailored data structures and associated methods into a user-friendly application. The PDE manages the definition of problems and supervises their solution. As opposed to the traditional architecture, where process engineers interact directly with the process modeling environment (PME), in the proposed IMCPSS architecture engineers interact solely with the PDE. Ultimately, the end-user does not require prior knowledge on the particulars of general-purpose modeling software or the mathematical concepts associated with the solution of advanced model-based problems. Therefore, the IMCPSS represents a valuable tool for process industries, as the facilitation of problem formulation is translated into minimization of time consumed in the analysis of processes, incorporation of plant data in less error-prone manner, and exploitation of synergy among activities based on process models.

3. Online parameter estimation, data reconciliation and transition planning

3.1 The process

The process used test the IMCPSS' online capabilities corresponds to a pilot-plant-scaled packed distillation unit available in the chemical engineering's unit operations laboratory at Louisiana State University. The distillation tower is 3 inches in diameter and has two packed sections, each of which contains 2' 8" of PROPAK 0.24" protruted stainless steel packing. The unit is fed with a multicomponent mixture of methanol, isopropanol, and water with weight percentages of 50, 30, and 20 respectively.

The unit is equipped with a total condenser, a partial reboiler, and a pump-back reflux system. Feed is introduced between the two packed sections and reflux enters at the top of the tower. An ethylene glycol system provides heat to a thermosiphon reboiler. Temperature, level, and flow rate controls are provided by the Honeywell Experion PKS control system. Liquid samples from feed, condensate, and bottoms can be taken manually and analyzed using gas chromatography. Figure 2 shows the process schematics. The first letter of a tag name indicates the type of measurement provided: temperature, flow rate, level percentage, and pressure, while the number identifies the loop.

The distillation tower, condenser, reflux drum, bottom sump, and reboiler were modeled considering the following assumptions:

- The packing can be represented by sections in which exiting streams are in equilibrium with each other. 22 packing sections were used.
- Murphy efficiencies are valid.
- Liquid and vapor holdups in each section are negligible. However, liquid holdups are considered for reboiler and reflux drum.
- Density and activity coefficients vary with column height. The UNIQUAC model was used to calculate activity coefficients.
- Pressure drop is negligible for the conditions tested during the validation of the IMCPSS.

The resulting model in gPROMS® comprehends 2995 variables (12 differentials). The index of the problem is one, so it was reduced using the available option in gPROMS. The IMCPSS interacted with the OPC server via the Automation Wrapper provided by the OPC foundation.

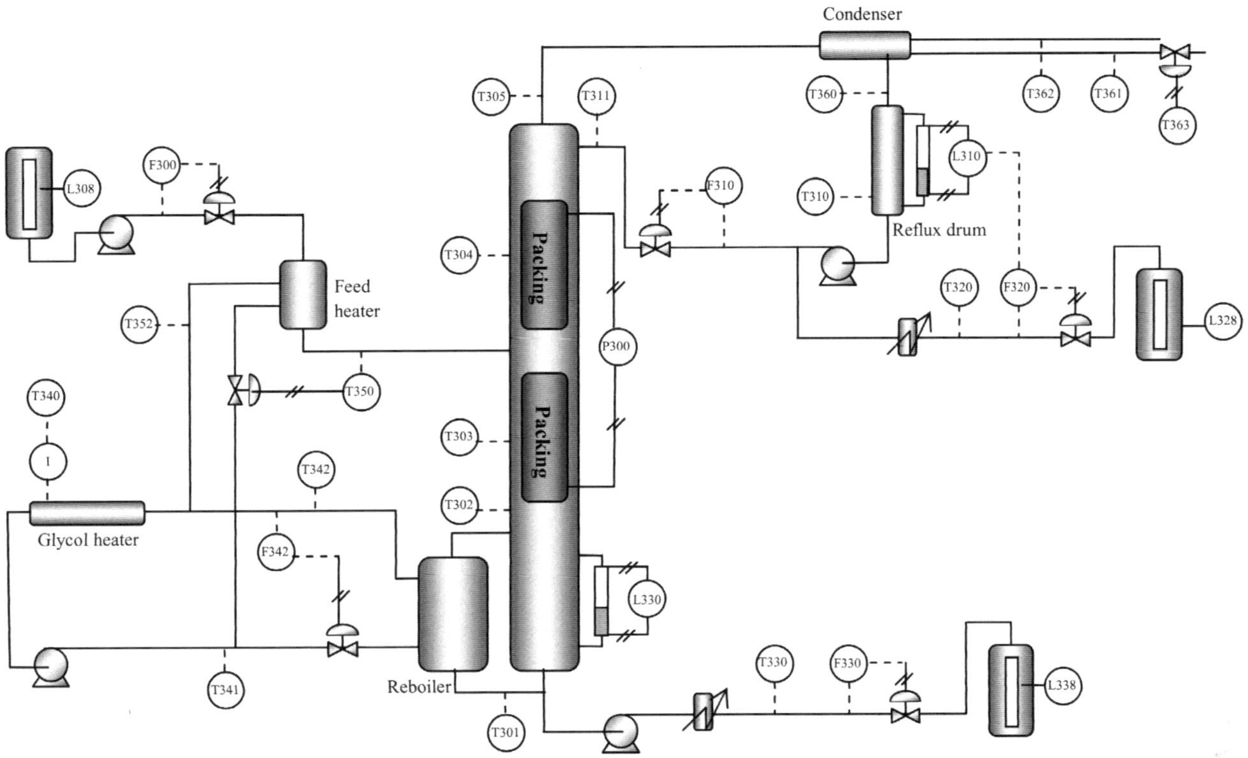

Figure 2. Packed distillation column schematics

3.2 Online estimation

The capability of the problem definition environment to formulate online problems is validated in this Section. Furthermore, the proposed procedures for online update in gPROMS are evaluated. The history horizon was set to 10 min and data is to be collected every 1min for a period of 60 min. Furthermore, we employed the GT-based objective function (Wang and Romagnoli, 2003; Aragon et al., 2008) with p=2, q=100 and errors bounded by $\pm 3\sigma$. Standard deviations were calculated from previous observations. After the distillation column reached an initial steady-state, online estimation was started with the history horizon and error bounds mentioned above. A step change of +20% in F310 was introduced 17 min after the online activity started.

3.2.1 Online parameter estimation/data reconciliation

Condenser overall heat transfer coefficient was estimated during reconciliation around the condenser (input sensors T361.PV, T340.PV and F342.PV; and output sensors T362.PV, T360.PV, T305.PV and T342.PV)‡. F310.PV was used as a forcing input to ensure the correct dynamic response of the system during the estimation. The measurement bias for T305 was also included as a decision variable. History horizon of 10, sampling time of 1min and total duration of 60min were used. The total number of decision variables is 32 (10 errors for each of three input variables, plus one bias and one process parameter).

The average solution time for a window was 340.90s, close to 6 times greater than the sampling interval, indicating that the reconciled trajectory was always delayed. Naturally, the number of reconciling sensors selected and computational power available is reflected on the time consumed by the reconciliation. Naturally, sensors to

‡ The .PV after the sensor name identifies the "process variable".

be reconciled should be selected carefully not only to provide sufficient redundancy but also to avoid unnecessary computational load, especially in process industries, where will certainly be hundreds or thousands number of decision variables.

Figure 3 show the reconciled trajectory of sensors T305.PV, T361.PV, and T360.PV, correspondingly. The estimated values of overall heat transfer coefficient and systematic error of T305 recognized three clear zones as shown in Figure 4a. Before the step change was introduced the average values were UA_c=10535.08 J/min.K and a bias in T305 of 3.081°F. During the transitional phase, these values ranged from 10527.9 to 11327.8 J/min.K for condenser UA_c and from 3.1428 to 4.6037 °F for the bias in T305. Then, the estimates averaged 11387.87 J/min.K and 4.1342°F correspondingly, once a new steady state was reached.

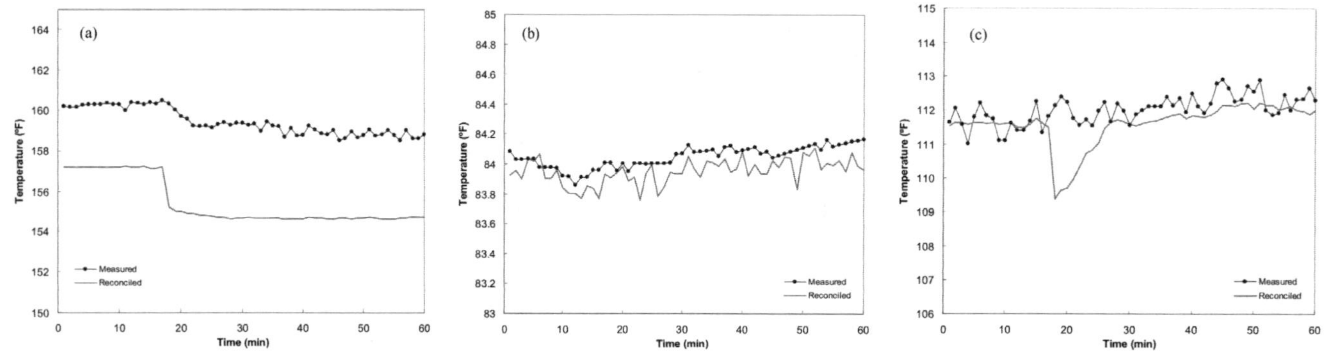

Figure 3: Observed vs. reconciled trajectory. (a) T305.PV (b) T361.PV (c) T360.PV (d) T362.PV

Just for comparison purposes, joint parameter estimation/data reconciliation was performed offline with the entire data set. The values obtained were UA_c=11315 J/min.K and T305.BIAS=4.257°F. It appears estimates tend to be those for higher times. Figure 4b shows that the reconciled trajectory of T360.PV does not follow properly the measured data. This suggests that executing online estimation leads to improved estimates. On the other hand, the computation time for online estimation was 17385.72s while it was 6287.66s for offline estimation. The benefits obtained with online estimation might outweigh the larger solution times, as shown for the case of T360.PV.

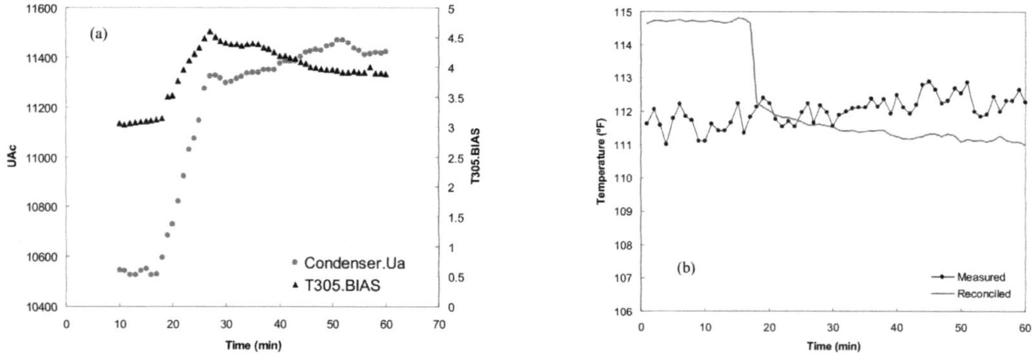

Figure 4: (a) Estimated bias in T305 and overall heat transfer coefficient in condenser UAc. (b) Observed vs. offline-reconciled trajectory of T360.PV

3.2.2 Transition planning

Under suboptimal operation, the transition from an operating point OP1 to the optimum operating conditions would proceed by performing single step changes on sensors F300, F310 and T340. However, this strategy might not constitute the optimal transition between those two operating conditions. Transition planning, a special case of dynamic optimization, is employed to find the optimal trajectory of these sensors to move from one operating point to another. This optimization procedure will be conducted to find the profiles of F300.PV, F310.PV and T340.PV that minimize variability in separation extent of methanol ensuring that the optimum throughput is achieved. The transition must be completed in five intervals with duration of 60min each.

The objective function that minimizes the separation extent variability is defined as:

$$z(\cdot)=\int_0^{t_f}\left|SE(t)-SE_{tgt}\right|dt \quad (1)$$

$$SE(t)=\frac{1-x_{MeOH,Bottom}}{x_{MeOH,Bottom}}\times\frac{x_{MeOH,Top}}{1-x_{MeOH,Top}} \quad (2)$$

Where $SE(t)$ and SE_{tgt} are the separation extent and its target value, respectively. The optimum profiles are presented in Table 1. The zero and last intervals are fixed by the economic optimum obtained previously by steady-state optimization.

Table 1: Transition planning to an economic optimum

Interval #	F300 (GPH)	F310 (GPH)	T340 (ºF)
0	2.143	3.2	207.555
1	2.637	3.942	215.048
2	2.757	4.150	216.591
3	3.724	4.139	215.239
4	2.683	4.111	214.769
5	3.106	4.2	216.853

Figure 5 shows the trajectory of the separation extent under suboptimum and optimum transition. Because the operating point is fixed for the last interval, the separation extent is also fixed. Nonetheless, it is clear from the picture that separation extent is maintained closer to the target value for the entire transition period when profiles given by the transition planning procedure are implemented.

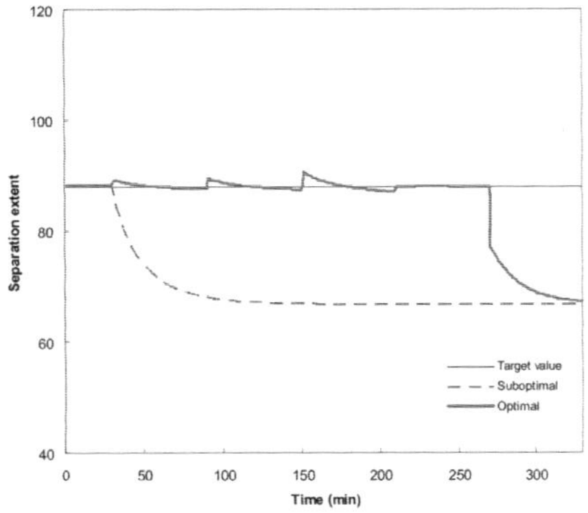

Figure 5: Simulated separation extent during transition to economic optimum

The distillation column was run first with the suboptimal operating conditions (i.e. one step change in manipulated sensors) and with the optimal profiles from transition planning. Figure 6 shows the setpoint profiles of manipulated variable T340.PV and the output variable F320.PV. Figure 6b indicates that minimizing the variability of the separation extent comes at expense of the distillate flow rate (F320.PV). While the suboptimum reaches the target production soon after the step changes have taken place, following the optimal recipe causes swings, with values of F320.PV as low as 0.8GPH.

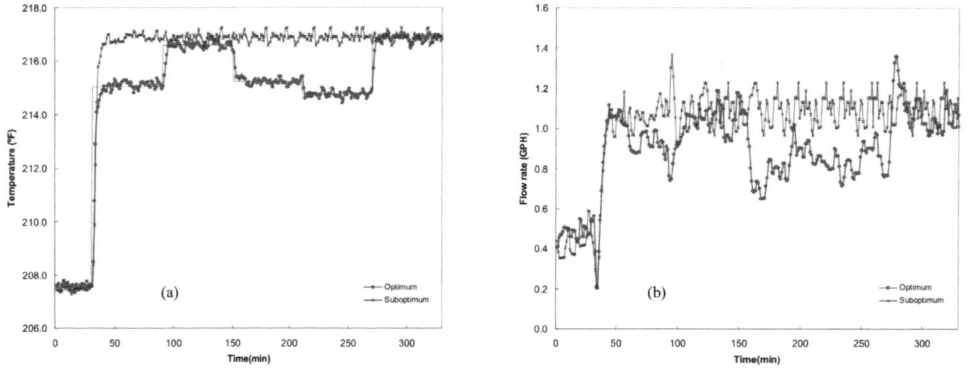

Figure 6: Setpoint profiles for transition to economic optimum. (a) T340.PV (b) F320.PV

4. Conclusion

The integrated model-centric process support system (IMCPSS) was validated for online parameter estimation/data reconciliation and transition planning by means of a simulated case consisting in a pilot-plant-scaled packed distillation column. The results indicate the viability of the IMCPSS to be used to formulate and solve online dynamic model-based activities of relatively high complexity when coupled with a powerful PME such as gPROMS. The case study presented in this work has corroborated the benefits that process industries can obtain with the application of model-based activities and the exploitation of synergistic relations between them.

This work represents a remarkable progress towards the development of tools capable of managing large and complex model-based problems in industrial environment. Future efforts should address the standardization interfaces for communication between PME and IMCPSS, in the same way CAPE-OPEN has promoted standardization of interfaces for communication between PME and PMCs. This will create a broader spectrum of problems and processes that can benefit from tools like the IMCPSS.

5. References

Al-Arfaj, M. Shortcut data reconciliation technique: Development and industrial application, AICHE Journal, 2006, 52, 141-417.

Aragon, D.; Rolandi, P.A. & Romagnoli, J. (2008). Robust (adaptive) dynamic data reconciliation and gross error estimation for unknown error distribution based on the generalized T distribution. 18th European symposium on computer aided process engineering, Lyon, France. June 1-4, 2008.

Aragon, D.; Rolandi, P.A. & Romagnoli, J. Implementation and validation of a model-centric support system within a pilot-plant scaled packed distillation column. AIChE national spring meeting. New Orleans, U.S.A. April 6-11, 2008.

Bai, S.; McLean, D. & Thibault, J. Autoassociative Neural Networks for Robust Dynamic Data Reconciliation, AIChE Journal, 2007, 52, 438-448.

Bezzo, F.; Muradore, R. & Barolo, M. Using structured and unstructured estimators for distillation units: a critical comparison, Computer Aided Chemical Engineering, 2005, 20, 1201-1206.

Bhat, S. A. & Saraf, D. N. Steady-state identification, gross error detection, and data reconciliation for industrial process units, Ind. Eng. Chem. Res., 2004, 43, 4323-433.

Espinosa, J. & Marchetti, J. Conceptual Modeling and Referential Control Applied to Batch Distillations. Ind. Eng. Chem. Res. 2007, 46, 6000-6009.

Rolandi, P.A. & Romagnoli, J.A. Integrated model-centric framework for support of manufacturing operations. Part I: The framework. Computers & Chemical Engineering, 2010, 34, 17-35.

Wang, D. & Romagnoli, J. A. (2003), A framework for robust data reconciliation based on a generalized objective function, *Ind. Chem. Eng. Res.* **42**, 3075-3084.

Packed Column Hydraulics and Continuous Phase Transitions

Brian Hanley, Aspen Technology, Inc., Burlington, MA USA

Introduction

Currently, no public domain correlation for packed bed hydraulics adequately predicts the pressure drop and flood point for random or structured packings over the full spectrum of vapor and liquid loadings, component physical properties, and column operating conditions normally encountered in industrial practice. Pressure drop is often a critical parameter in column design – for example, in separations involving polymerizable or thermally sensitive components – while flood determines the maximum operational capacity of a column. With this in mind we explore the hypothesis that the flooding locus in a packed column operated countercurrently represents an envelope of second order phase transitions from vapor to liquid continuous operation. We show that an appropriate renormalization of column pressure drop data to the incipient flood point for the particular liquid flowrates examined results in a collapse of these data onto a single "master" curve. This data collapse occurs for virtually all packing types, sizes, and materials of construction. A description of the data reduction procedure will be given. In addition, we will show several examples for specific packings. Further, we will show how these results can be extended to systems with physical properties other than those of the original experimental system.

Pressure Drop as an Example of a Continuous Phase Transition

This paper will expand on the hypothesis presented in earlier articles by Hanley, et al., that the flooding locus in a packed column represents an envelope of continuous phase transitions from gas continuous to liquid continuous operation (or vice versa). Therefore, as with other types of continuous phase transitions, renormalizing packing pressure drop/vapor load/liquid load data to the appropriate flood points should result in a collapse of those data onto a single master curve whose abscissa is fractional approach to flood, F_S/F_S^* (or C_S/C_S^*). Further, since physical laws cannot depend on arbitrarily chosen units of measurement the pressure drop per unit height itself must be renormalized properly. Two potential choices for the appropriate renormalizing factor are the hydrostatic liquid pressure drop per unit height, $\rho_L g$, and the dry pressure drop per unit height for the packing at the same F_S as that for the wet pressure drop, $\Delta p_{dry}(F_S)/Z$. We will show that the first choice is the correct one. The resultant scaled ordinate becomes $[(\Delta p/Z)]_{2\phi}/\rho_L g$. This result has immediate consequences. For example since the pressure drop at flood for water is often taken to be 2" H_2O/ft (~16 mbar/m), then the pressure drop at flood for a liquid with a different density will be $2(\rho_L/\rho_W)$ "H_2O/ft ($16[\rho_L/\rho_W]$ mbar/m). In short, then, we are proposing a transformation of data from a form like that of equation (1)

$$\left(\frac{\Delta p}{Z}\right)_{2\phi} = P\left(F_S, C_L, \nu, \sigma, \ldots\right) \tag{1}$$

into a dimensionally consistent scaled form given by equation (2)

$$\frac{\left(\frac{\Delta p}{Z}\right)_{2\phi}}{\rho_L g} = S\left(\frac{C_S}{C_S^*(C_L, \nu, \sigma, ...)}\right) \tag{2}$$

Data Analysis

We expect the rescaled two-phase pressure drop to be only a function of the fractional approach to flood. In the figure below, the selected four points should map to a single point when plotted against fractional approach to flood.

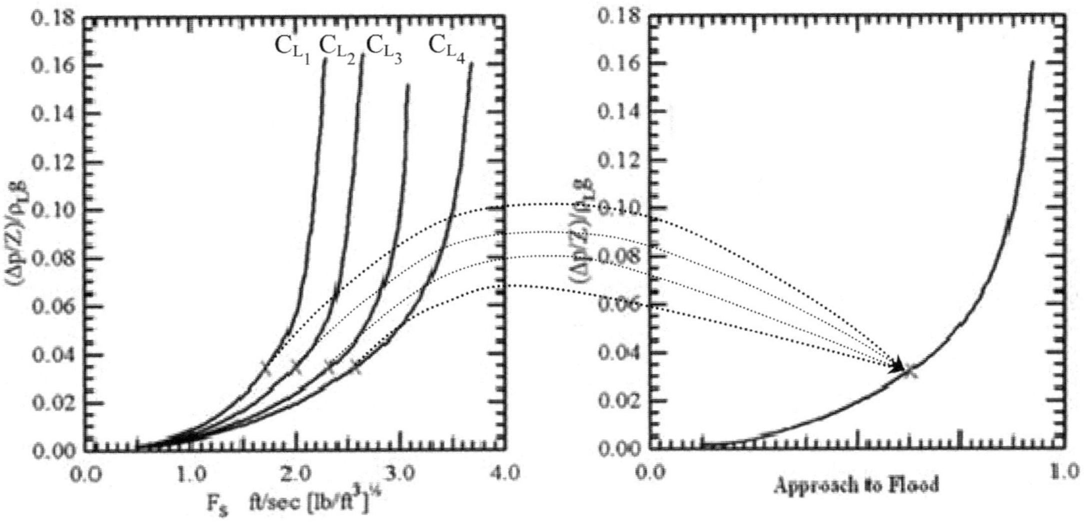

Figure 1: Schematic showing the mapping of [$(\Delta p/Z)_{2\phi}/\rho_L g$, F_S, C_L] triplets onto a single (($\Delta p/Z)_{2\phi}/\rho_L g$, f) point, where "f" stands for approach to flood at constant liquid load.

The same mapping should occur for any set of four points at constant pressure drop. This observation leads to the following strategy for finding flooding velocities: flooding velocities are those which minimize the total variance (or total standard deviation) in the calculated fractions of flood among all sets of points with constant pressure drop. For a single set of four points, we would choose the flooding rates - F^*_{Sa}, F^*_{Sb}, F^*_{Sc}, F^*_{Sd} – so that they minimize the expression for the standard deviation, σ:

$$x_a = \frac{F_{Sa}}{F^*_{Sa}} \quad x_b = \frac{F_{Sb}}{F^*_{Sb}} \quad x_c = \frac{F_{Sc}}{F^*_{Sc}} \quad x_d = \frac{F_{Sd}}{F^*_{Sd}} \tag{3}$$

$$\langle x \rangle = (x_a + x_b + x_c + x_d)/4 \tag{4}$$

$$\sigma = \sqrt{\frac{(x_a - \langle x \rangle)^2 + (x_b - \langle x \rangle)^2 + (x_c - \langle x \rangle)^2 + (x_d - \langle x \rangle)^2}{3}} \tag{5}$$

The minimization procedure for this example would be carried out for all sets of four points at constant pressure drop. At least one flooding velocity needs to be fixed at the start of the fit since any multiple of the actual flood velocities will also collapse the data. Least squares can be used to perform the optimization by setting the objective function to be zero for all points (we want the standard deviation or variance to be as close to zero as possible).

The procedure as outlined above assumes that constant pressure drop data across several liquid loads are readily available. In terms of raw data this will rarely be true – experiments are usually not performed in a mode wherein pressure drops are held constant while vapor load is varied at constant liquid load. Instead, most experiments hold the liquid load constant while scanning across the vapor load. Pressure drop is the measured variable. The required constant pressure drop data can be estimated in a number of ways. For example, smooth curves can be drawn through the data and then points at constant pressure drop read off. Alternatively, one could incorporate some type of numerical interpolation procedure (linear, cubic spline, or smoothing spline, for example) into the least squares routine.

Results

We have carried out the procedure described above for air/water data on approximately 200 individual packings, including both random and structured types. In the great majority of cases the data collapse as described above quite well. Figures 2a through 2f below demonstrate that the data collapse occurs for a variety of different packing types across a range of different vendors. Cases do occur, however, where the proposed collapse does not seem to happen. Figures 2g and 2h, for IMTP 40 and FLEXIPAC 1YHC (manufactured by Koch-Glitsch, LP), are examples. It is unclear why certain datasets fail to collapse as expected. It is possible, given the difficulty in performing these types of experiments well, that the datasets themselves are simply in error, perhaps because of flowmeter and/or pressure transducer miscalibrations or because of water in the tap lines leading to the pressure transducer diaphragms.

Extensions to Systems Other Than Air/Water

The same type of data collapse observed for air/water would be expected to occur for any gas/liquid combination used to carry out this type of experiment. The absolute values of the gas rates at flood, however, would change primarily due to changes in the physical properties of the liquids used in the experiments. Therefore, in order to be able to predict the pressure drop in systems other than air/water we require a method for adjusting the flood points measured in air/water to values that would occur in a system with different physical properties.

The air/water column flooding data can be fit well with a Wallis-type of equation[1]

$$\sqrt{C_S^*} + m_W \sqrt{C_L^*} = c_W \qquad (6)$$

where C_L^* is the superficial liquid velocity at flood, C_S^* is the density corrected superficial vapor velocity at flood, and m_W and c_W are fitting parameters.

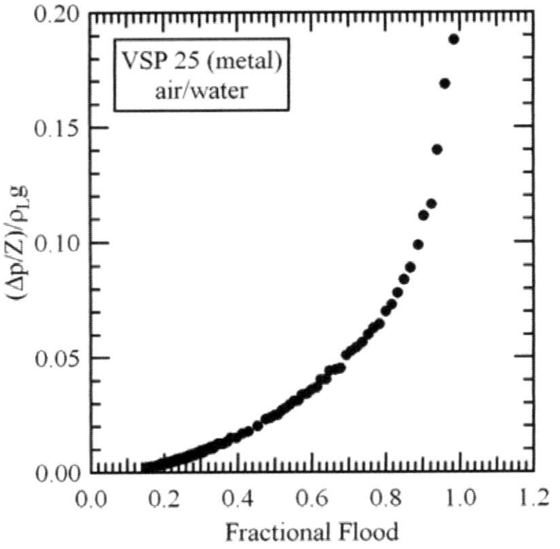

(2a)

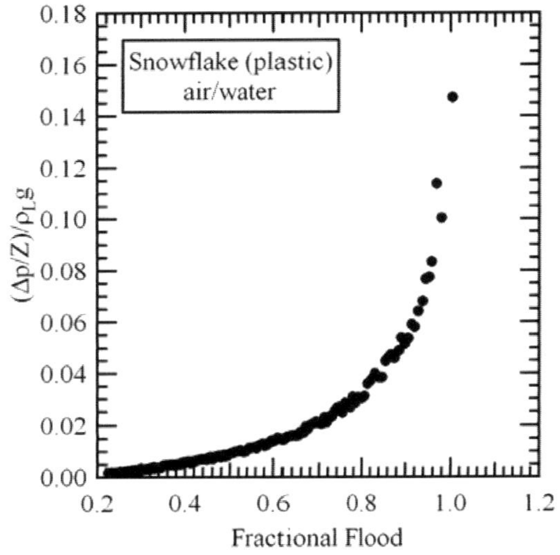

(2b)

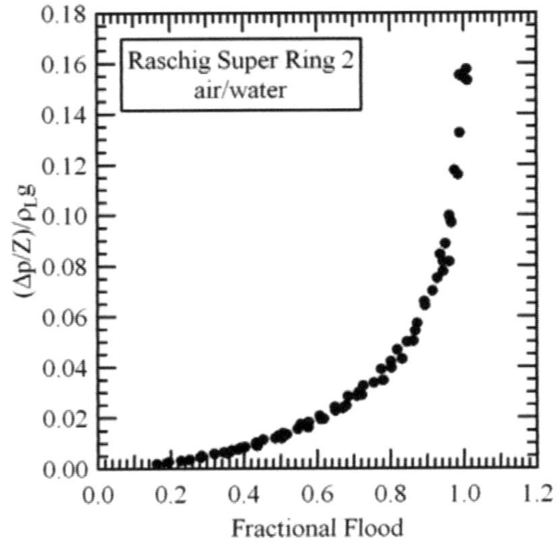

(2c)

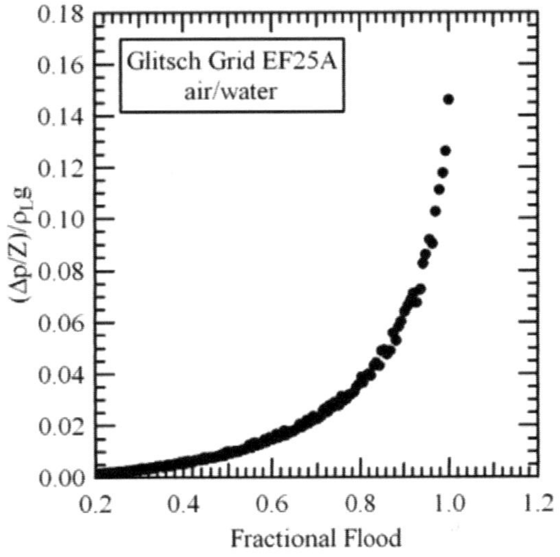

(2d)

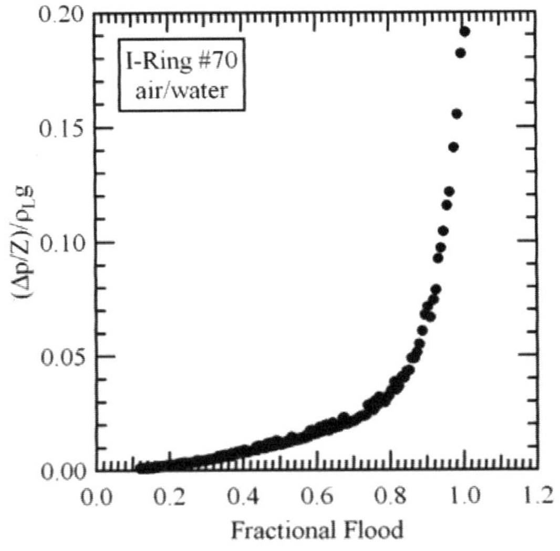

(2e)

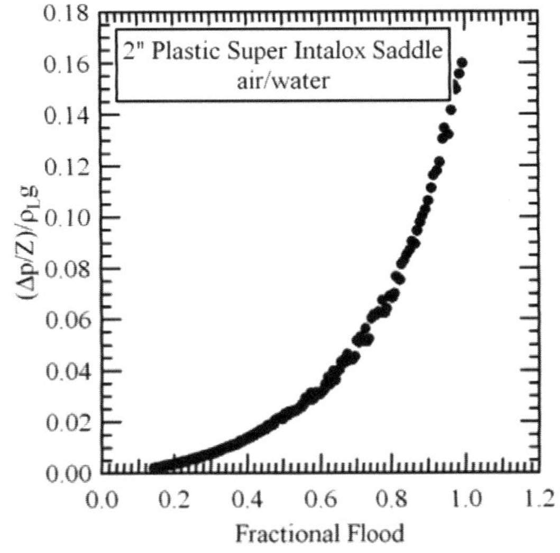

(2f)

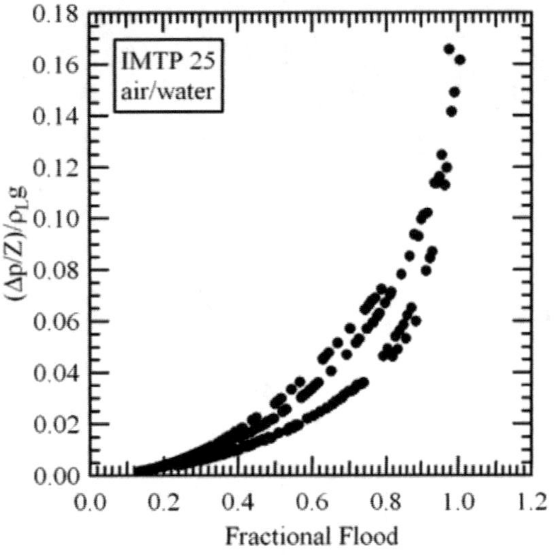

(2g)

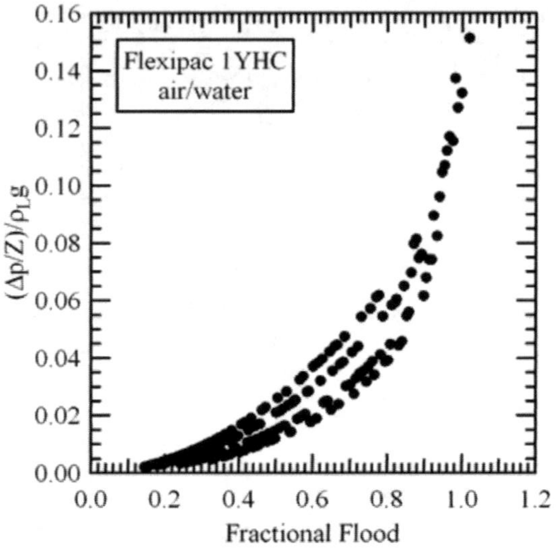

(2h)

Figure 2: Data collapse: a) VFF metal VSP #25[2], b) Koch-Glitsch plastic Intalox Snowflake packing[3], c) Raschig Super Ring No. 2[4], d) Koch-Glitsch Grid EF25A[5], e) Sulzer I-Ring #70[6], f) Koch-Glitsch 2" plastic Super Intalox Saddle[2], g) Koch-Glitsch metal IMTP 40[7], and h) Koch-Glitsch FLEXIPAC 1YHC[8]

Hanley[9] showed that

$$m = \mathbf{M}\left(\frac{d_e^2 \rho_L g}{\sigma}, \frac{d_H^3 \rho_L^2 g}{\eta_L^2}\right) = \mathbf{M}(\text{Bo}, \text{Ga}) \qquad c = \mathbf{C}\left(\frac{d_e^2 \rho_L g}{\sigma}, \frac{d_H^3 \rho_L^2 g}{\eta_L^2}\right) = \mathbf{C}(\text{Bo}, \text{Ga}) \tag{7}$$

where "Bo" is the Bond number and "Ga" is the Galilei number. Because the functional forms for the dependence of the Wallis parameters on the Bond and Galilei number are unknown, let us assume for the sake of simplicity that the Wallis equation parameters are power laws of the Bond and Galilei numbers

$$m \propto (\text{Ga})^a (\text{Bo})^b \qquad c \propto (\text{Ga})^x (\text{Bo})^y \tag{8}$$

Two limiting cases arise: 1) the Galilei number dominates flooding, or 2) the Bond number dominates. In each limiting case we are able to make a remarkable observation – the effects of the liquid's physical properties on the flood locus can be quantified by observing the effects of the equivalent diameter on the Wallis parameters "c_W" and "m_W" for a geometrically similar packing family using air/water data only.

Figure 3 presents an example showing the variation in the Wallis parameters c_W and m_W with the equivalent diameter of the packing. The power law exponents, a and b or x and y, are easily determined for case 1 or case 2 above. We have examined flood point data for a number or non-aqueous systems. In Figure 4 the percent difference between the reported flood points and calculated flood points based on the dimensionless Galilei and Bond number corrections are presented. Figure 4a demonstrates that the Galilei number (which involves the liquid's viscosity) fails to adequately account for liquid physical property effects when the Galilei number is small. Figure 4b, on the other hand, shows that reasonable flood point physical property corrections can be made to air/water flood data for the complete range of Bond numbers (involving the liquid's surface tension) covered by these data.

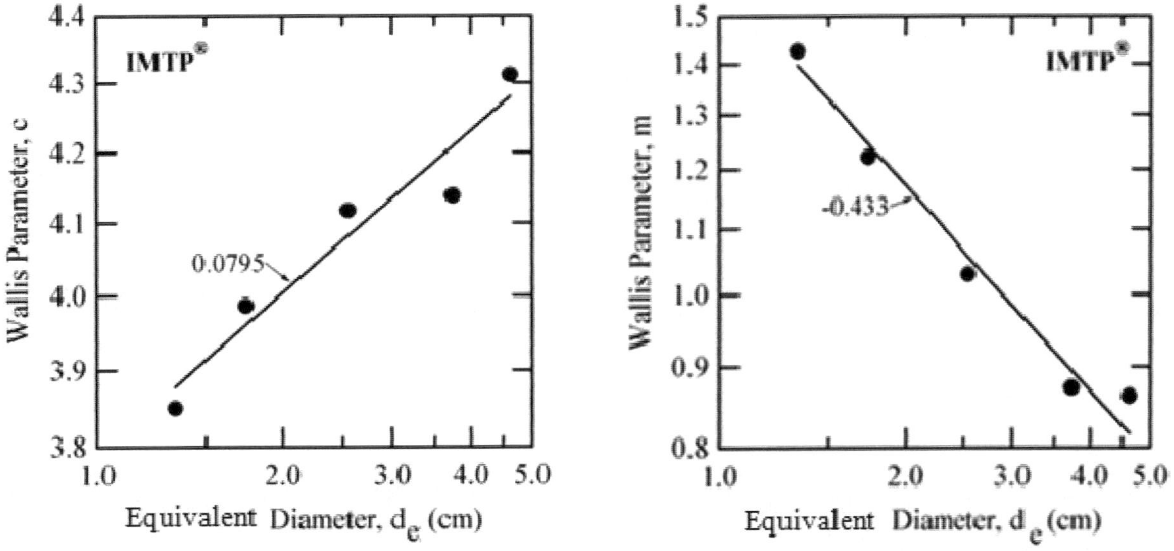

Figure 3: Variation of the air/water Wallis constants with equivalent diameter for IMTP random packing family.

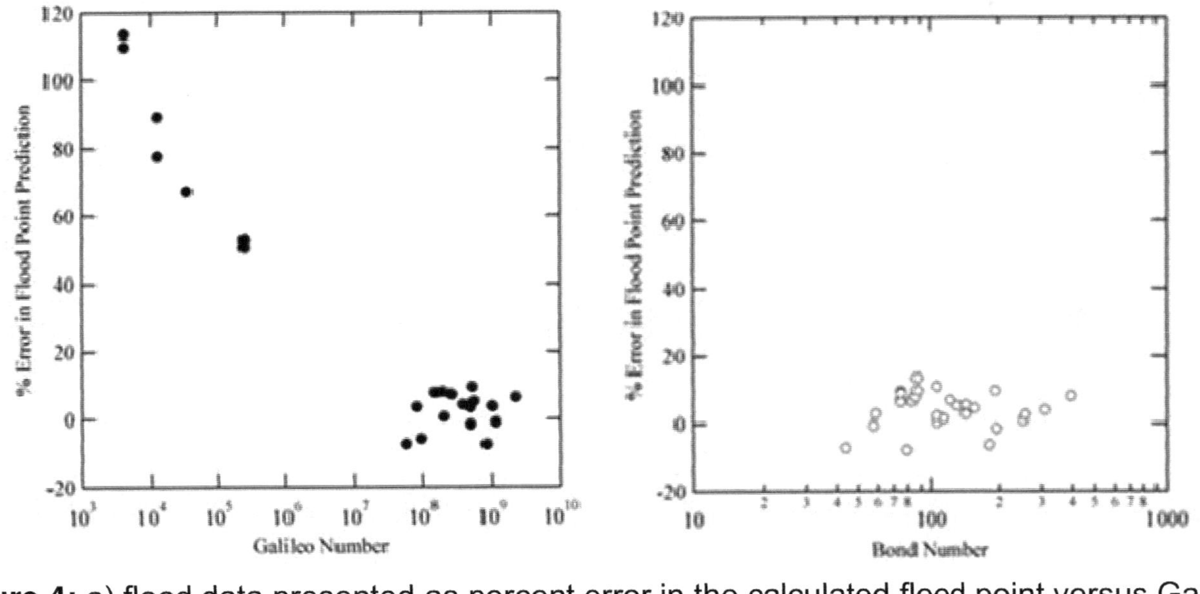

Figure 4: a) flood data presented as percent error in the calculated flood point versus Galileo number, and b) the same data displayed as percent error in the calculated flood point versus Bond number.

Comparison to Experiment for Non-Aqueous Systems

Figure 5 is a comparison of pressure drop predictions for three different packings in three different chemical systems with experimental data taken on these same packings and systems. It is important to note that the original two-phase pressure drop data on which these predictions were made were taken in air/water. Clearly the pressure drop predictions are in line with experiment. Further, the capacity of the columns in these experiments has also been adequately captured.

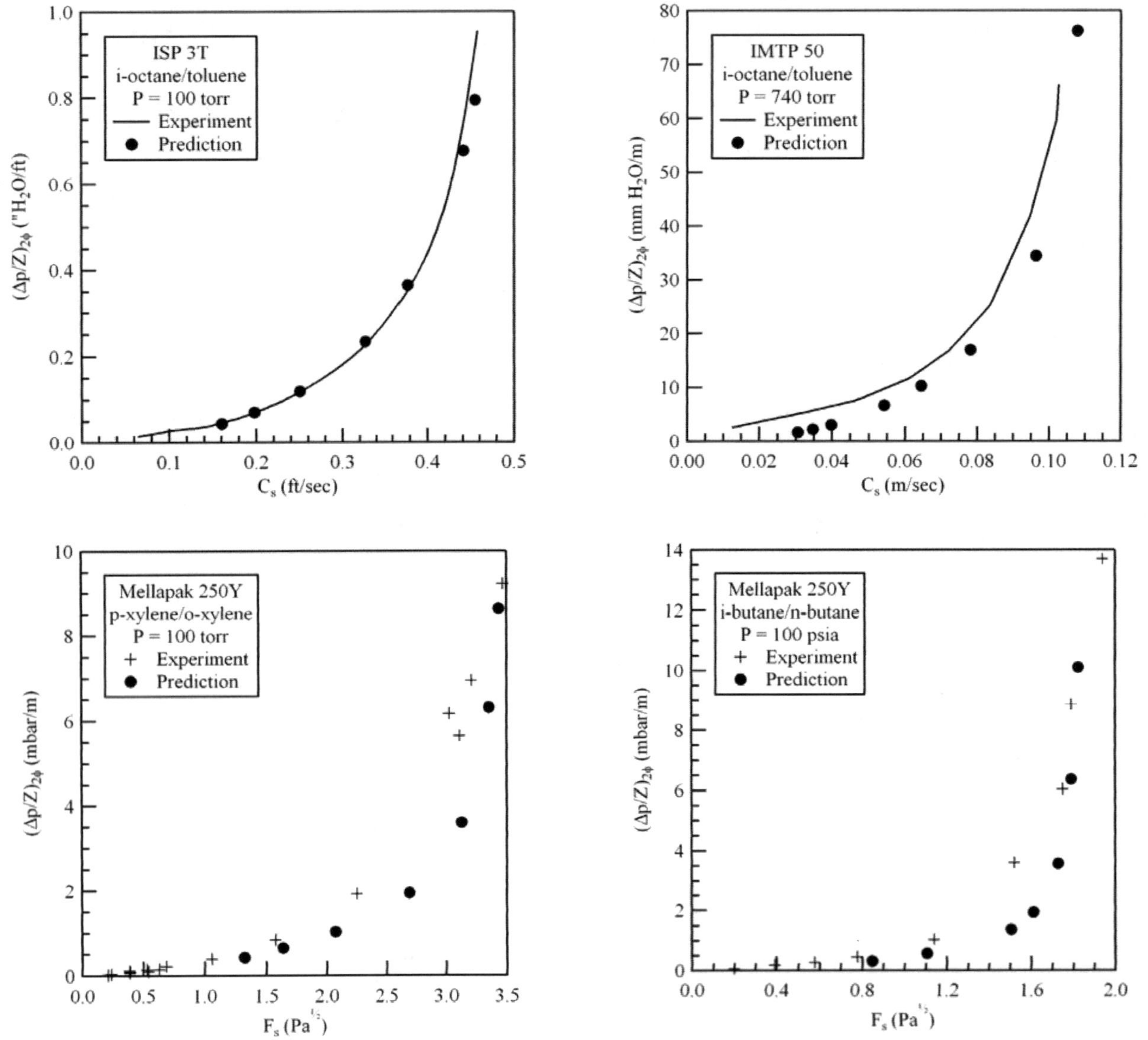

Figure 5: Comparison of model predictions to experiment for: a) i-octane/toluene at 100 torr for ISP 3T structured packing[10], b) i-octane/toluene at 740 torr for IMTP 50 random packing[11], c) p-xylene/o-xylene at 100 torr for MELLAPAK 250Y structured packing[12], c) i-butane/n-butane at 100 psia for MELLAPAK 250Y structured packing[12].

Conclusion

A two-phase pressure drop prediction method for packings has been developed. It is based on the observation that packed column flooding is an example of a continuous phase transition and that, therefore, we can expect a level of universal behavior in the pressure drop/approach to flood regardless of the actual vapor and liquid loads involved. Predictions by this method are superior to those made via other literature models because they capture both the magnitude of the pressure drop as well as the approach to flood for the column without need for empirical correction. The success of the data renormalization procedure described in this paper is further proof of the crucial role played by the continuous phase transition commonly referred to as flooding in packed column hydraulics.

References

[1] K.J. McNulty and C.L. Hsieh, AIChE Annual Meeting, Los Angeles, CA, November, 1982

[2] Vereinigte Füllkörper-Fabriken GmbH & Co. KG Bulletin

[3] Koch-Glitsch, LP Bulletin KGPP-1 2M050EE, 2003

[4] M. Schultes, 2001 Spring AIChE Meeting, Houston, TX, April, 2001

[5] Koch-Glitsch, LP Bulletin KGSS-1 2M0503E, 2003

[6] Sulzer Chemtech Bulletin "Metal Random Packing"

[7] Koch-Glitsch, LP Bulletin KGIMTP-2 2M1303E, 2003

[8] Koch-Glitsch, LP Bulletin FPHC-3M0600B, 2000

[9] B. Hanley, et al., AIChE Annual Meeting, Reno, NV, November, 2001

[10] Norton Chemical Process Products Corp. Bulletin ISP-2, 1994

[11] http://www.cheresources.com/distillationmodel4.shtml

[12] C. Fitz et al., *Ind.Eng.Chem.Res.* **38**(1999) 512-518

(84e) A Dimensionless Model for Predicting the Mass Transfer Area of Structured Packing

Robert E. Tsai, Chemical Engineering, The University of Texas at Austin, Austin, TX

A. Frank Seibert, Separations Research Program, The University of Texas at Austin, Austin, TX

R. Bruce Eldridge, Chemical Engineering, The University of Texas at Austin, Austin, TX

Gary T. Rochelle, Chemical Engineering, The University of Texas at Austin, Austin, TX

The mass transfer area of nine structured packings was measured in a 0.427 m ID column via absorption of CO_2 from air into 0.1 kmol/m^3 NaOH. The area was most strongly related to packing size (125-500 m^2/m^3) and liquid load (2.5-75 $m^3/m^2 \cdot h$). Surface tension (30-72 mN/m) had a weaker but significant effect. Gas velocity (0.6-2.3 m/s), liquid viscosity (1-15 mPa·s), and flow channel configuration had essentially no impact on the mass transfer area. Surface texture (embossing) increased the effective area by 10% at most. The ratio of mass transfer area to specific area (a_e/a_p) was correlated within limits of ±13% for the entire experimental database:

$$\frac{a_e}{a_p} = 1.339\left[\left(\frac{\rho_L}{\sigma}\right) g^{1/3} \left(\frac{Q}{L_p}\right)^{4/3}\right]^{0.116}$$

See more of this Session: Advances in Distillation & Absorption III
See more of this Group/Topical: Topical 8: Distillation Topical

SUITABILITY OF HIGH PERFORMANCE RANDOM AND STRUCTURED PACKINGS FOR CO2-ABSORPTION IN POWER PLANTS

Michael Schultes, Simon Chambers, Brad Fleming
Raschig GmbH-Jaeger Technologies
Mundenheimerstrasse 100, 67061 Ludwigshafen, Germany
1611 Peachleaf St., Houston TX, USA 77039

Abstract

To meet the target of optimized CO_2-Capture performance from Power Plant flue gas, high capacity/high mass-transfer-efficiency/low pressure drop modern packings are required. With a wide variety of high performance packings available on the market, a series of test campaigns were required to screen those most suitable for CO_2-removal. To this effect, hydraulics-efficiency research from Total Reflux Distillation tests in the Commercial/Pilot Test Columns at Fractionation Research Incorporated (F.R.I.) and at the Separations Research Program (SRP) was performed. Additional research to measure pressure drop with the air-water system and mass transfer measurements from CO_2 absorption supplement the findings from distillation testing. Test results show distinct performance advantages of modern high performance metal random and structured packings over the current state-of-the-art high performance packings. The data presented will validate the necessity of such test campaigns in the selection process of packings for CO_2 Absorption /Stripping processes. The packings involved in the comparison studies are Raschig Super-Ring® random packings, various sized Pall Rings and Third Generation random packings; Raschig Super-Pak® structured packings, Sulzer Mellapak® including Mellapak-Plus®, and the Montz B1® packing types.

Keywords: CO2-Capture, Maximum Efficient Capacity, Effective Surface Area, Pressure Drop

(84g) Simulation of Interfacial Rayleigh-Bénard Convection in the Gas-Liquid Mass Transfer Process Using a Lattice Boltzmann Method

Fu Bo, Department of Chemical Engineering, State Key Laboratory of Chemical Engineering, Tianjin University, Tianjin, China

Yuan Xi-Gang, Department of Chemical Engineering, State Key Laboratory of Chemical Engineering, Tianjin University, Tianjin, China

Liu Bo-Tan, Department of Chemical Engineering, State Key Laboratory of Chemical Engineering, Tianjin University, Tianjin, China

Chen Shu-Yong, Department of Chemical Engineering, State Key Laboratory of Chemical Engineering, Tianjin University, Tianjin, China

Abstract

Interfacial convection caused by the interfacial density gradient produced by concentration gradient in interphase mass transfer process is cited as interfacial Rayleigh-B¨¦nard (IRB for short) convection. The Rayleigh-B¨¦nard convection at the interface displays an important role on mass transfer rate. Therefore, it is highly important to study the IRB convection for many engineering applications such as absorption, extraction, distillation, etc.. A large amount of experimental work and theoretical analysis have devoted to obtain observations of the structures of the IRB convection, measurements of enhanced mass transfer rate and the linear stability analysis of Rayleigh-B¨¦nard problem. In this paper, by introducing a mass transfer model of fixed disturbance(FDM for short), both flow and concentration fields of the two-dimensional IRB convection were simulated by the lattice Boltzmann method (LBM) at mesoscopic level with two-distribution model in the gas-liquid mass transfer process of CO_2 absorption into quiescent ethanol. The main purposes of simulation are to qualitatively compare the IRB convection patterns with the images taken by Schlieren technique and analyze the relationship between the IRB convection patterns and mass transfer rate and quantitatively analyze the relationship between liquid-phase mass transfer coefficient and number of fixed disturbance points on the interphase. The lattice Boltzmann method, of which the basic concept is to incorporate the physics of the problem into a simplified kinetic equation such that the correct macroscopic behavior of the fluid is recovered, have been widely used to study the Rayleigh-B¨¦nard convection phenomenon caused by heat transfer. However, to the author's knowledge, this is the first time the LBM approach is applied to investigate the Rayleigh-B¨¦nard convection caused by the mass transfer. In the present study, a series of disturbance points are set on the gas-liquid interphase with the assumption that once the convection structure is formed it continues at the fixed local position and never fade until the whole process is over.

Considering the simplicity and accuracy, D2Q9 model of LBM is chosen for the simulation. In the D2Q9 model, also known as the two-dimensional nine-velocity model, the physical space is discretized into a square lattice. The corresponding particles at each node have 9 moving directions. In gas-liquid system of CO_2 absorption into quiescent ethanol, we only concern a two-dimensional unsteady and isothermal mass transfer of a vertical section of the liquid in the simulator.

At the beginning, pure ethanol was confined in a rectangle box and its free upper boundary contacted with CO_2. The mass transport of CO_2 absorption into quiescent ethanol took place through the upper boundary. The absorption system was Rayleigh-B¨¦nard unstable (*Ra*>0) and Marangoni stable (Ma<0). Therefore, the IRB convection appeared in liquid-phase if the density gradient induced by the diffusion of the solute into a vertical section of the liquid bulk was beyond a critical value. Considered of the lateral and lower wall effects, the inner space dimension of our simulator was 0.01m°Á0.02m, which the liquid thickness was selected 0.01m. The simulations were carried out on a uniform 30°Á60 grid. Physical properties of pure ethanol and its saturated CO_2 solution in absorption process for our LBM simulation are listed in Table 1.

Table 1 Physical properties of pure ethanol and saturated CO_2 solution at *T*=293.2 K, *p*=1 atm

¦Ñ$_s$ kg°¤m^{-3}	¦Ä¦Ñ kg°¤m^{-3}	C_{Sat} kg°¤m^{-3}	D m^2°¤s^{-1}	μ mPa°¤s	c_s m°¤s^{-1}	Signs of *Ra*	Signs of *Ma*
788.9	+1.1	5.15	4°Á10^{-9}	1.15	1168	+	-

Here ¦Ñ is the density of pure ethanol; ¦Ä¦Ñ is the differences between the properties of the pure ethanol and the saturated CO_2 solution; C_{Sat} is the concentration of saturated CO_2 solution and c_s is speed of sound in ethanol.

IRB convection phenomena is simulated for situations in which the number of fixed disturbance existing and uniformly distributed on the interface of CO_2 and ethanol are 1,2,3,4,5,9,11,14,19,29 and 59.

The case for *n*=5 and *t*=1s, 3s, 6s, 10s, the transient contours of CO_2 concentration are shown in Fig. 1.

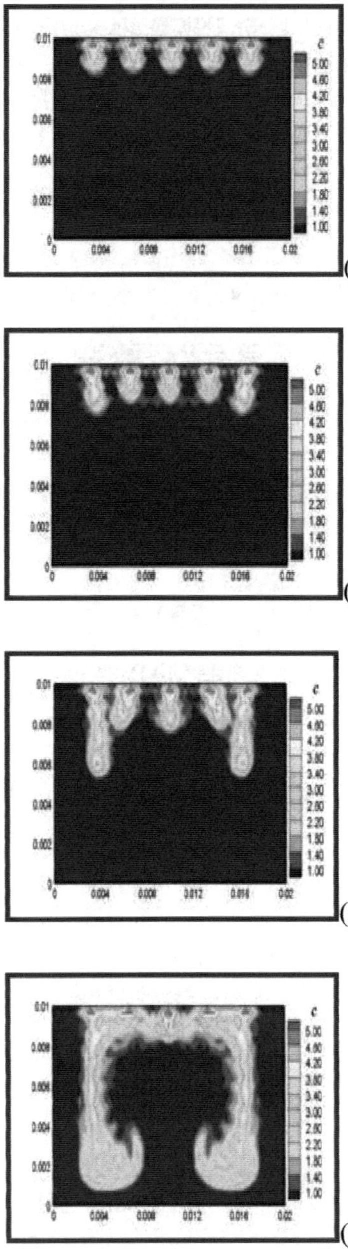

(a) t=1 s

(b) t=3 s

(c) t=6 s

(d) t=10 s

Fig. 1 Transient contours of the CO_2 concentration for different t at n=5

From Fig. 1, at the beginning of mass transfer, the IRB convection appeared on the fixed disturbance points, and extended quickly to the whole interface; the interfacial convection intensity increased with the time. The critical contact time for the onset of the IRB convection was found to be less than 1s. Then the liquid with higher concentration began to stream downward in a plume density-driven convection pattern and reached the bottom of the bulk liquid accompanying by upward backflow with swing back and forth at t=10s. At last, two inverse-mushroom convective structures formed. It should be noted that there were mutual induction effects between two adjacent fixed disturbance points, and the mutual induction effect played an important role in forming the plume density-driven convection pattern in the mass transfer process.

When $n \leq 5$, heavy superficial liquid with nearly saturated with CO_2, descended in a downward plume density-driven convection pattern, when $n \geq 9$, the plumes drifted from the middle of the concentration contours to the sides and converged on the bottom, which was termed as the diverging flow pattern that the layers of nearly saturated liquid descended along rectangle lateral walls induced by wall effect; interfacial turbulence such as chaotic cells and short hook-like plumes seen in vicinity interface were not clear in concentration contours but charts of velocity vector.

The liquid-phase mass transfer coefficients with different number of fixed disturbance points were calculated by the concentration distributions for gas-liquid contact time 7s in mass transfer process of CO_2 absorption into quiescent ethanol as shown in Fig. 2.

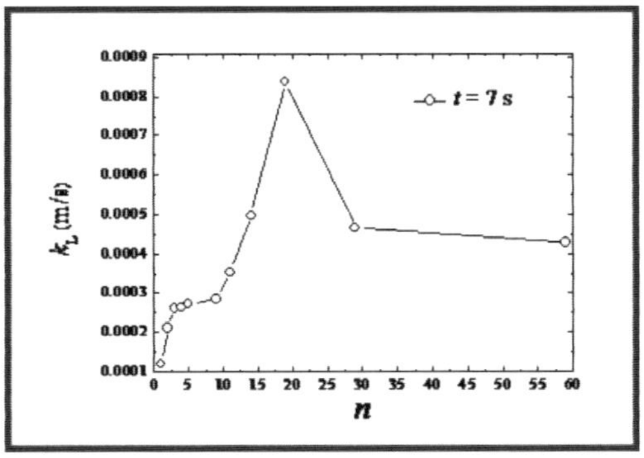

Fig. 2 The liquid-phase mass transfer coefficient variation with number of fixed disturbance points

The liquid-phase mass transfer coefficient increases at first then decreases to a nearly stable value with the increase in number of disturbance points as shown in Fig. 2. And it can be seen that the curves for the liquid-phase mass transfer coefficient can be divided into two regions: (1) an increasing region with $n \leq 5$, where the mass transfer was controlled by the plume density-driven convection pattern; (2) an increasing at first then decreasing region with $n \geq 9$, where the mass transfer was controlled by two diverging flow patterns and number of interfacial turbulence patterns such as chaotic cells and short hook-like plumes. In the first region, k_L first increases markedly then slightly increases with the increase in n, and k_L is proportional to the plumes convection intensity increased nonlinearly with the increase in n. It is noteworthy that the compound plumes at n=4, n=5 are combined with single plume density-driven convection pattern by mutual induction effects between two adjacent fixed disturbance points, and the compound plume convection intensity is larger than single plume convection intensity. In the second region, k_L first increases considerably then decreases markedly to a nearly stable value with the increase in n that can be explained that: when $n \geq 9$ and $n \leq 19$, the diverging flow intensity and the number of interfacial turbulence patterns rapidly increases in spite of the decreasing interfacial turbulence intensity with the increase in n; when $n \geq 19$ and $n \leq 29$, the diverging flow intensity and number of short hook-like plumes convection intensity rapidly decreases in spite of the increasing number of interfacial turbulence patterns with the increase in n; when $n \geq 29$, although the diverging flow intensity decreases, the mass transfer intensity induced by molecular diffusion increases with the increase in n. Moreover, the transient stage of convection pattern where k_L slightly increases from $n=5$ to $n=9$ shown in Fig. 2 is that two compound plume density-driven convection patterns converted into two diverging flow patterns with nearly equal intensity. Consequently, the mass transfer rate depends on number of plumes and types of patterns in the IRB convection.

The present study is a heuristic investigation which attempt to relate the mass transfer rate with the types and intensity of the IRB convection structures by varying the number of disturbance points. We hope this result could be helpful to uncover the mechanism of interphase mass transfer and further work is needed in the future.

See more of this Session: Advances in Distillation & Absorption III
See more of this Group/Topical: Topical 8: Distillation Topical

(109a) FRI Experimental Unit - Evaluation and Upgrade

A.L. Parker, Fractionation Research, Inc., Stillwater, OK

M.R. Resetarits, Fractionation Research, Inc., Stillwater, OK

A.Y. Ogundeji, Fractionation Research, Inc., Stillwater, OK

D.E. Williams, Fractionation Research, Inc., Stillwater, OK

T.J. Cai, Fractionation Research, Inc., Stillwater, OK

The experimental Unit of Fractionation Research Inc is located in Stillwater, Oklahoma, and includes a low-pressure and a high-pressure distillation column. Each has its own reboiler and its own condenser. Piping links the two reboilers; other piping links the condensers. For the last 15 years, neither column could be operated above a reboiler duty of 17 MMBtu/hr. Over the last two years, most of the equipment items comprising the Unit were evaluated. No significant safety concerns were found. Process bottlenecks and inefficiences were identified. Capital spending addressed both. It is now possible to run each column at a reboiler duty of 23 MMBtu/hr. The next generation of trays and packings can be evaluated - once they are invented. The final cost of the upgrade was far less than what was originally budgeted.

See more of this Session: Advances in Distillation & Absorption IV

See more of this Group/Topical: Topical 8: Distillation Topical

(109b) Challenges in Fundamental Valve Tray Efficiency Modeling in the Mixed-Froth Regime

Anand N. Vennavelli, Chemical Engineering, Oklahoma State University, Stillwater, OK

Dadasaheb B. Salunke, Chemical Engineering, Oklahoma State University, Stillwater, OK

James R. Whiteley, Chemical Engineering, Oklahoma State University, Stillwater, OK

The froth regime is the most likely flow regime on an industrial distillation tray (Kister 1992). It has been the flow regime of interest for most tray efficiency models. The froth regime is classified into two sub-regimes (Hofhuis and Zuiderweg 1979, Lockett 1986) – bubbling-froth and mixed-froth. The two-phase hydraulics and mass-transfer characteristics of the bubbling-froth are well understood and have been implemented in several tray efficiency models. The mixed-froth, however, is relatively less understood and has only been modeled on sieve trays. The recent advances in fundamental modeling of sieve tray efficiency (Garcia and Fair 2000, Syeda et al. 2007) can be attributed to improved mixed-froth modeling. Current mixed-froth models, however, have several empirical aspects that limit generalization. In this paper, the challenges in the fundamental modeling of the mixed-froth are discussed. Empirical and theoretical aspects of the current mixed-froth models and their applicability to valve trays are evaluated. This work is part of the ongoing work at OSU to identify the similarities, differences, and the gaps in sieve and valve tray efficiency models.

See more of this Session: Advances in Distillation & Absorption IV

See more of this Group/Topical: Topical 8: Distillation Topical

(109c) Design of a 4-Product Dividing Wall Column: One Vs Multiple Partition Wall Configurations

Zarko Olujic, Process & Energy, Delft University of Technology, Delft, Netherlands

I. Dejanovic, University of Zagreb, Zagreb, Croatia

Lj. Matijasevic, University of Zagreb, Zagreb, Croatia

Ivar J. Halvorsen, Ict, SINTEF, Trondheim, Norway

Sigurd Skogestad, Department of Chemical Engineering, Norwegian University of Science and Technology, Trondheim, Norway

Helmut Jansen, Julius Montz GmbH, Hilden, Germany

B Kaibel, Julius Montz GmbH, Hilden, Germany

Preliminary evaluations using a simple but reliable short-cut method indicated that a 15 component aromatic's mixture can be separated very efficiently into four fractions according to given product specifications employing a complex dividing wall column (DWC). Using predicted vapor and liquid splits as initial guesses facilitated carrying out rigorous simulations which indicated that a three-columns-in-parallel internal configuration, which implies installation of two partition walls in parallel, enables higher energy efficiency than more practical, one partition wall configuration. The later has been implemented in practice (one column in operation at BASF) as a packed column equipped with high performance structured packings provided with efficient and robust wall scrappers, which allow fixing the off-centre partition walls by packed beds. From installation and troubleshooting point of view, un-fixed partition walls that are assembled form detachable elements require minimum installation/ dismantling effort and time. In this paper a further step has been made, i.e. available technology used to dimension properly a DWC with multiple partition walls placed in parallel in off-centre positions to accommodate internal flow as required according to outcome of rigorous simulations. The liquid splits can be easily set by mechanical means while the multiple vapor splits are arranged by adjusting the parallel sections pressure drop accordingly. This can be achieved by choosing proper design of structured packing and accompanying internals. Simple and complex design options of the four-product DWC are presented, including related total annualized costs, and relative advantages/disadvantages discussed. Although the energy saving gain is substantial the configuration is that complex that from process and mechanical point of view that increased effort and capital as well as related uncertainties do not justify practical implementation.

See more of this Session: Advances in Distillation & Absorption IV
See more of this Group/Topical: Topical 8: Distillation Topical

(109d) Random Packing Surface Area Influence On Hydraulic-Efficiency Performance: Challenging Conventional Wisdom

Simon Chambers, JAEGER PRODUCTS INC., Bedford, TX

Brad Fleming, JAEGER PRODUCTS INC., Bedford, TX

Michael Schultes, Raschig Ring Department, RASCHIG GMBH, Ludwigshafen, Germany

It is well known that the surface area of random packings directly influences their hydraulic-efficiency performance. In particular the lower the surface area, the higher the throughput capacity combined with lower pressure drop but at the expense of a decreased theoretical stage count (high HETP). The converse is true as the surface area increases. These heuristics were validated following a 2 year research program at FRI in the mid 1980's, in which four Second Generation Pall Rings were tested.

Nowadays there are numerous third generation ring-saddle hybrids and the first fourth generation random packing (the Raschig Super-Ring) available on the market. Such advancements in packing geometries raise the question of whether the above mentioned heuristic still applies. It was put to the test by comparing test results obtained from the SRP 430mm ID Distillation Pilot column of four random packings spanning three generations.

The recently tested Raschig Super-Ring No. 0.5 of Surface Area 250 m2/m3 was compared against two Second Generation 25 and 50 mm Pall Rings (207 and 102 m2/m3 surface areas) and a Third Generation 40 mm Ring-Saddle of Surface Area 170 m2/m3. The data presented will show that the relative high surface area Raschig Super-Ring No. 0.5 exhibits higher capacities combined with as good as or lower pressure drops compared to the lower surface area Second and third Generation types with much better efficiencies. On the whole the fluid dynamic behaviour within the packing element and not just surface area alone should be considered when evaluating the performance of random packings.

See more of this Session: Advances in Distillation & Absorption IV
See more of this Group/Topical: Topical 8: Distillation Topical

(109e) Improved Improved Mass Transfer Correlations for Random and Structured Packings

Brian Hanley, Aspen Technology, Inc., Burlington, MA

Hiren Shethna, Process Modeling, Aspen Technology, Inc., Burlington, MA

The reliability of rate-based calculations for packed columns operated countercurrently is known to be limited by the accuracy of the correlations used to estimate the mass transfer coefficients and the interfacial area (Švandová, et al., 2008; Adler, 2000). Typically the mass transfer coefficient matrices for multicomponent systems are based upon generalizations of mass transfer coefficient expressions developed from data on binary systems. Clearly, the successful analysis of experimental data for binary systems is crucial to the acceptance of rate-based modeling as an improvement over the traditional equilibrium stage model for multicomponent systems.

This paper explores the development and limitations of today's public domain mass transfer correlations for packed columns. First, we shall consider the types of data generally available to the modeler. Next, we demonstrate that traditional attempts to decompose binary HETP data into fits for a vapor-side and a liquid-side mass transfer coefficient equation along with an equation for the mass transfer area are, at best, accurate to within a multiplicative constant. This finding implies that other types of experiments are needed to unambiguously define the vapor-side and liquid-side contributions to the total resistance. We also demonstrate that it is sometimes not possible to find a unique set of fitting parameters from efficiency data alone; either theoretical arguments are needed to set some parameters or, once again, some other type of experiment is needed to resolve the impasse. Our studies further reveal that physical property models do play a substantial part in the success or failure of any particular mass transfer correlation and that care is therefore warranted in their selection.

Finally, we present new mass transfer correlations for Pall ring style metal random packings, IMTP style metal random packings, and Mellapak "Y" style metal structured packings. The development of these correlations will be discussed, their limitations examined, and their performance relative to several public domain mass transfer correlations presented.

Z. Švandová, et al., Chem. Eng. J., 140 (2008) 381–390. Adler, S., et al., Vision 2020: 2000 Separations Roadmap, AIChE Center for Waste Reductions and U.S. DOE, 2000.

See more of this Session: Advances in Distillation & Absorption IV
See more of this Group/Topical: Topical 8: Distillation Topical

Advanced Simulation of Structured Packing via CFD

Scott A. Owens* and R. Bruce Eldridge.
Separations Research Program, University of Texas, Austin, Texas 78712
* To whom all correspondence should be addressed. Tel.: (512) 471-1087. Fax: (512) 471-1720. E-mail: sao293@che.utexas.edu.

INTRODUCTION

The Navier-Stokes equations can be applied to turbulent flows with the addition of terms to account for the average effects of turbulence.[1] The modified equations are known as the Reynolds-Averaged Navier-Stokes (RANS) equations and can be solved iteratively in time and space by applying numerical methods. Computational Fluid Dynamics (CFD) refers to the use of computers to solve these equations.

The addition of turbulent terms in the RANS equations requires additional equations for closure. These terms come from well documented empirical correlations such as the k-ε or Reynolds Stress turbulence models.[1] Because closure is obtained empirically, the accuracy of CFD simulations must be thoroughly investigated prior to their use as a predictive tool. This process is known as validation and typically consists of blind comparisons between simulation predictions and experimental data.

Recent advances in multi-core processors and parallel processing allow CFD researchers to study flows previously inaccessible due to the complexity and/or size of the geometry. Because CFD solves the equations of motion, momentum and energy it can yield far more detailed data than analogous experimentation which is often limited by instrumentation. In these ways, CFD promises to be more economic, faster, and safer than comparable experimental campaigns.

METHOD

CFD simulations are widely employed in automotive, marine, and aerospace design. In these industries, validation can be achieved by comparing CFD predicted streamlines to smoke trails which are visually tracked across the geometry surface. By contrast, most chemical operations lack optical access and are designed to mix or disperse fluids. This has prevented the acquisition of data to validate models and prevented the widespread acceptance of CFD in the chemical industry.

Recent research within the PSTC explored the use of X-ray based particle tracking to overcome the need for optical access.[2] The researchers tracked neutrally buoyant, X-ray absorbing particles entrained in water flowing through structured packing. The particles' positions in each frame were obtained directly. Local velocities were calculated by combing this information with the frame rate and macroscopic streamlines were obtained by summing each frame along a particles path. These data will be compared to the velocity distribution and streamlines predicted by CFD simulation.

Additional validation data will be obtained from pressure drop experiments. Measurements will be made with packed section identical to that used in the work of Owens *et al*. Single phase pressure loss will be studied using first nitrogen then water as a bulk phase flowing upward through the packing. The nitrogen experiments will be conduct at F_S between 0.5 and 2.5 ft/s*$[lb/ft^3]^{1/2}$. Experiments utilizing water as a bulk phase will be run at the Reynolds numbers which are equivalent to the nitrogen experiments. Pressure drop will be measured through a) three-half elements, b) four-half elements, and c) two-whole elements of Sulzer M250Y. Data taken with three-half elements of MP250.Y will be compared to predictions made by CFD simulations. The data taken in four-half and two-whole elements of MP250.Y will be used to confirm and quantify the joint effects observed in Owens *et al.*[2]

CFD simulation requires the creation of a digital representation of the geometry to be modeled. This is traditionally accomplished via Computer Aided Drafting (CAD). However, CAD geometry generation often begins by simplifying the geometry. These simplifications arise out of an inability to recreate fine features or a desire to reduce the time and cost associated with geometry creation.

This project utilizes X-ray CT scans to generate CFD geometry. CT scans yield a three-dimensional density map. This data is then converted to a stereo lithography (STL) file. STL files discard all data which do not describe a density interface. The resulting file is much smaller and directly importable into most CFD meshing programs. This method creates a geometry with fidelity limited only by the resolution of the scanner (typically sub-millimeter, see figure 1). It is expected that CT generated geometry will yield geometry more efficiently than CAD while also reducing error.

Figure 1: X-ray CT produces high fidelity geometry models as shown above.

The Star-CD software package STAR-CCM+ has been selected for this project. STAR-CCM+ will be used to adaptively generate tetrahedral meshes with a lower limit of 0.4 mm on a side. The mesh will be converted to a polyhedral mesh in order to reduce computational demand and increase accuracy.[1] Calculations and data analysis will be carried out on the SPUR and RANGER high performance computing clusters at the Texas Advanced Computing Center.[3] Turbulence simulation will utilize an improved k-ε model (i.e. Realizable or RNG) to model constant density, single-phase nitrogen flow.

RESULTS AND DISCUSSION

A single half-element of plastic Sulzer M250Y was scanned at the University of Texas Computed Tomography facility. The CT scan was performed with 16-bit accuracy, a slice thickness of 0.3125 mm, and took approximately five hours to complete. The scan data was then replicated, rotated, and stacked twice to generate the complete stacked height (figure 2). A wall and inlet/outlet planes were added to form a "water-tight" geometry.

Figure 2: The assembled packing geometry.

The geometry was then imported into STAR-CCM+ with a vertex merge tolerance of 0.05 mm. The Surface Wrapper, Surface Remesher, and Volume Meshing sub routines were run with a global mesh minimum of 0.45 mm. Curvature based refinement was used with a requirement that any discernable gap contain at least one mesh element. The resulting tetrahedral mesh was comprised of more than 50 million mesh elements (figure 3, left). By converting to a polyhedral mesh this was reduced to less than 30 million elements (figure 3, right).

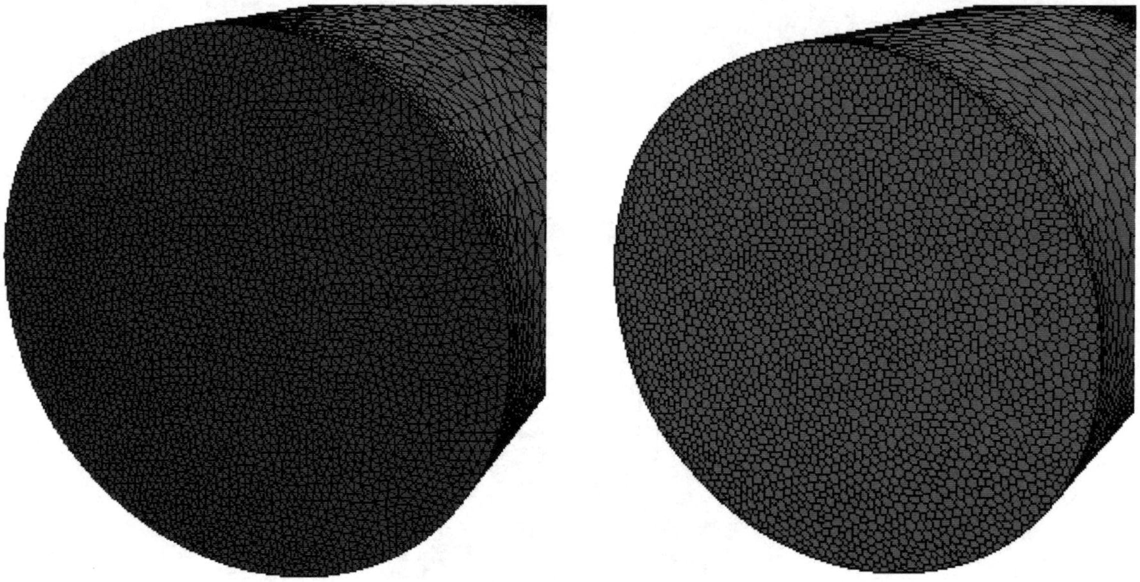

Figure 3: A tetrahedral (left) and polyhedral (right) mesh are shown on the pressure outlet and wall boundaries of the CFD geometry.

Initial simulation was conducted on the SPUR visualization cluster. Turbulence was calculated with the Realizable k-ε two-layer model. The velocity and pressure profiles were analyzed after initialization. The bulk velocity was shown to be slightly below the 1 m/s boundary condition. By contrast, flow near the wall was much higher (figure 4, top). This wall flow is likely a result of the relatively large wall-gap present in the lab scale packing as compared to industrial scaled packing. The pressure plot (figure 4, bottom) also showed interesting results. There is a very visible step change between the area immediately below and above the wiper bands. This is expected as the wiper band represents a flow constriction. A similar but smaller step correlates with the location of the packing joint. This suggests that there is elevated pressure loss at the joints relative to the mid packing zones. These results are consisted with and shed light on earlier research which documented increased hold-up at the joint under two phase flow.[4] This increased pressure loss could create a back pressure which would require a higher liquid hold up to overcome and maintain the liquid flow rate.

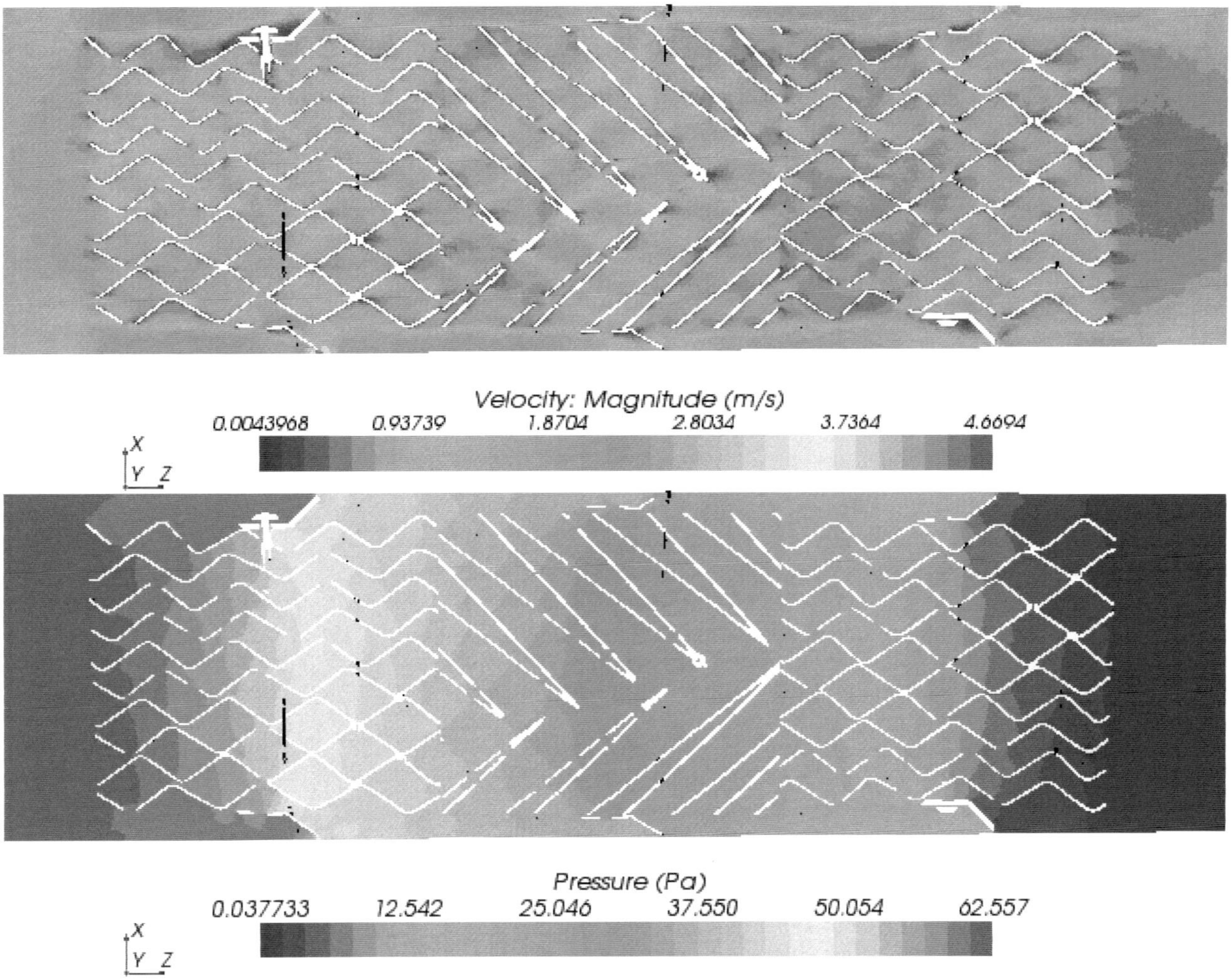

Figure 4: Velocity (top) and pressure (bottom) contour maps showing flow from left to right on a plane through the center of the geometry.

The results of the validation experiments are show in figure 5. The results predict a slightly lower pressure drop than similar experiments taken on the SRP air-water column.[5] This could be attributed to the wall flow which reflected in the CFD simulations. Data was also taken to evaluate the effect of joints on the pressure performance of structured packing. The results are show in figure 6. The data shows an increased pressure loss associated with a greater number of joints. This is in agreement with the results show by Owens *et al.*

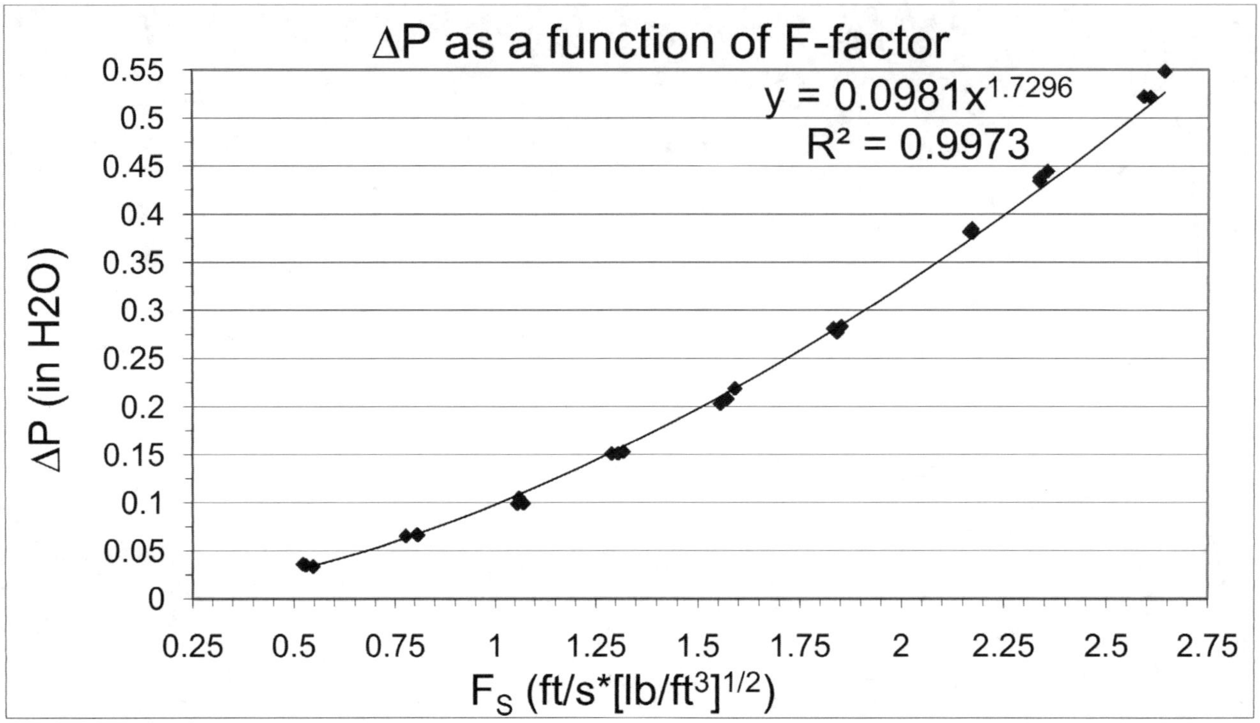

Figure 5: This graph shows data obtained from geometry identical to the CFD geometry.

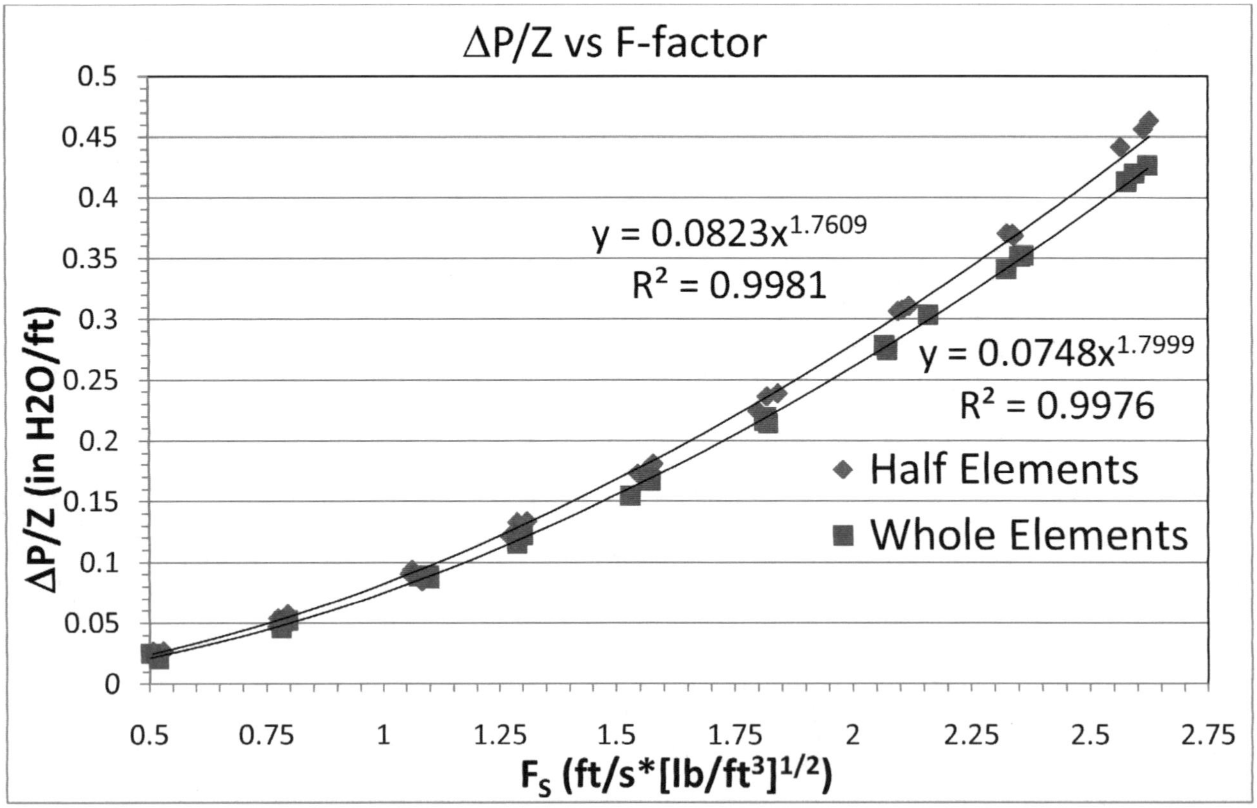

Figure 6: This graph shows experiments performed to quantify the effects of packing joints in two feet of packing.

References

1. Casey, M.; Wintergerste, T., *Best Practice Guidelines*. 1 ed.; European Research Community On Flow, Turbulence and Combustion: 2000; p 94.

2. Owens, S. A.; Kossmann, A.; Farone, J.; Eldridge, R. B.; Ketcham, R. A., Flow Field Visualization in Structured Packing Using Real Time X-ray Radiography. *Industrial & Engineering Chemistry Research* **2009,** 48, (7), 3606-3618.

3. The Texas Advanced Computing Center. www.tacc.utexas.edu

4. Green, C. W.; Farone, J.; Briley, J. K.; Eldridge, R. B.; Ketcham, R. A.; Nightingale, B., Novel Application of X-ray Computed Tomography: Determination of Gas/Liquid Contact Area and Liquid Holdup in Structured Packing. *Ind. Eng. Chem. Res.* **2007,** 46, (17), 5734-5753.

5. Tsai, R. E.; Schultheiss, P.; Kettner, A.; Lewis, J. C.; Seibert, A. F.; Eldridge, R. B.; Rochelle, G. T., Influence of Surface Tension on Effective Packing Area. *Industrial & Engineering Chemistry Research* **2008,** 47, (4), 1253-1260.

(133a) An Innovation in Advanced Mass Transfer Technology: The Combination of a High Performance Tray and a Novel Downcomer Distributor for Optimal Performance

Lindsey Vuong, Process Design, AMT International, Inc., Plano, TX

Kuang Wu, Process Design, AMT International, Inc., Plano, TX

Adam T. Lee, Process Design, AMT International, Inc., Plano, TX

Zainab Bt. Kayat, Research and Technology, Petronas Group Technology Solutions, Malaysia

Amy A. Yusof, Petronas Gas Berhad, Malaysia

The deviation from ideal flow pattern on standard commercial trays can be caused by the recycling of liquid on the tray or by the existence of stagnant regions of liquid. In order to eliminate these issues, a "Downcomer Distributor"[1] was invented and developed.

The Downcomer Distributor, along with a standard and another proprietary downcomer design, was tested in 2007. It was found experimentally that specific advantages are achieved with the Downcomer Distributor at a wide range of vapor and liquid flow rates, including the elimination of stagnant regions above the tray decks and the reduction of back-flow, thus enabling effective utilization of the entire tray deck for mass transfer. Compared with standard and other proprietary downcomer designs, the distribution of volumetric liquid flow across the tray deck is essentially similar for all paths along which that liquid flows. Thus, the Downcomer Distributor results in improved efficiency.

A successful commercial application of the high performance trays with Downcomer Distributors is field-proven with the de-bottleneck of a commercial-scale gas plant in June 2008. ADV Pinnacle Performance Trays with Downcomer Distributors were installed in both the LPG and Depropanizer Columns. Compared to the prior columns' performance, it was determined during the performance test run period that the columns after revamp are able to handle a greater throughput, the on-spec propane and butane product rates are greater, and thus the calculated incremental production rates significantly exceed the Project targeted incremental production rates. After revamp the tray efficiencies of the LPG and Depropanizer Columns were increased, which results in substantial improvements in column energy consumption, product purities and recoveries, and operation flexibility.

[1]The Downcomer Distributor is patent-pending.

KEYWORDS: downcomer distributor, high-performance tray, downcomer, mass transfer

See more of this Session: Advances in Distillation & Absorption V
See more of this Group/Topical: Topical 8: Distillation Topical

(133b) Liquid Phase Resistance to Mass Transfer in Structured Packing

John Billingham, Praxair, Inc., Tonawanda, NY

Kirk Larson, Praxair, Inc., Tonawanda, NY

John Ricotta, Praxair, Inc., Tonawanda, NY

G.X. Chen, Praxair, Inc., Tonawanda, NY

Abstract

Due to its low pressure drop and high mass transfer efficiency, structured packing becomes the most popular column internal in vacuum and low pressure distillation columns. Its application to the air separation industry has resulted in significant savings in energy consumption. Although many mass transfer efficiency models were developed by various research groups, the liquid phase resistance to mass transfer using structured packing in distillation is barely determined and little studied. This paper shows the significance of liquid phase resistance in predicting mass transfer efficiency, particularly for high purity systems. The predicted liquid phase resistance from various literature models1-3 are examined and compared. From limited literature data4, it was found that the liquid phase resistance to mass transfer is about 30% of the gas phase resistance for the C6/C7 system at 5 and 24 psia. The predicted liquid phase resistance from the Gualito model1 agrees well with experimental data. However, the Billet model2 significantly over predicts the liquid phase resistance. This discrepancy in the predicted liquid phase resistance from two models complicates the prediction of HETP as a function of the slope of the equilibrium line and the (L/V) ratio.

See more of this Session: Advances in Distillation & Absorption V
See more of this Group/Topical: Topical 8: Distillation Topical

Next Generation Scalable Models for Mass Transfer in Packed Column

K. Nandakumar[1] and Karl T. Chaung[2]

[1]Gordon A. and Mary Cain Chair Professor, Cain Department of Chemical Engineering
Louisiana State University, Baton Rouge, LA, 70803, USA
(e-mail: nandakumar@lsu.edu)

[2]Department of Chemical & Materials Engineering, University of Alberta,
Edmonton, T6G 2G6, CANADA

Abstract— In this talk we present our results on modeling the hydraulics and mass transfer in random and structured packed columns. Two distinctly different approaches are discussed. The first one is based on computational fluid dynamics wherein volume averaged equations for gas-liquid flow in a packed bed are used. The concept of volume averaging, its merits and limitations will be reviewed. This is done in the framework of interpenetrating continua and it requires closure relations to capture the interface transport processes such as drag and mass transfer. Hence predictions from such models must be validated against data from well controlled experiments. In our laboratory we measure the liquid distributions over Pall rings under various conditions and compare such data against CFD predictions. HETP data from Fractionation Research Inc. are used to validate mass transfer models. The main advantage of such an approach is that the model predictions can remain scale invariant over a certain range. Hence CFD models can help in minimizing scale up studies at the pilot plant scale.

In an alternate approach, we simulate the random packing of complex shaped packing elements like Pall rings or Super Raschig rings in a container using collision detection algorithms. Once the position and orientation of each packing element is determined, we can interrogate the data to get porosity and surface area distribution within the bed. Such data can then be used in the CFD simulations or cell based models. We have explored both and these results will be discussed. The same approach is also used for studying not only performance of existing structured packings, but also in the design of new structured packings.

Keywords-Computational fluid dynamics, multiphase flows, chemical process models.

I. Introduction

With the continuing advances in both the hardware and algorithms, computational tools have become an integral part of design and analysis of many engineering processes including chemical and mineral processes. In many operations ranging from chemical, biological and environmental processes, fluid mechanics plays a central role in influencing the advective and diffusive mixing, transport and reaction of chemicals and species. Even for a single phase flow, turbulence remains a formidable challenge and direct numerical simulation of turbulence can only shed light on the simplest of ideal flow situations and any realistic simulation of turbulence by direct numerical simulation places so much demand on computer simulation because of the wide range of spatial and temporal scales that must be resolved. Hence the Reynolds averaged Navier-Stokes equations is often the pragmatic tool that is used for addressing the effects of turbulence on other transport and reaction processes. Fortunately, a wide range of practical turbulence models have been developed tested and codified into commercial computational fluid dynamics packages, the $k-\epsilon$ *and* $k-\omega$ models, the Reynolds stress models, and LES models being examples of such efforts. These models provide some form of closure models for the information that has been lost in the time averaging process.

The problem is exacerbated much more, when one considers multiphase flows. Unfortunately the nature of most flows in chemical and mineral processing operations (such as distillation columns, packed or tray, packed and fluidized bed reactors, gravity separation vessels such as oil-water separators, liquid-liquid extraction columns etc) involve both dispersed phases of complicated structures and turbulence. In this work we review the current status of knowledge in this area and identify the future needs in making these tool a reliable one to understand and improve the performance of packed column operations involving multiphase flows. This presentation is in honor of Prof. K. T. Chuang and is based on 10 articles that we have co-authored with him in this area and summarizes the major findings from these studies. The reader is referred to the original articles [1-10] for details.

II. The Volume Averaged NSE

A. The concept

We restrict our attention to fluids that under normal, single phase flow situations, are described adequately by the incompressible, Navier-Stokes equations on a continuum scale. We further focus our attention on the dynamics of such gas-liquid systems flowing through a fixed solid phase (a packed column). The dispersed phase has a random distribution and the volume fraction of the dispersed phase is significant. Direct numerical simulation of such flows from the underlying Navier-Stokes equation remains a formidable challenge, although progress is being made in this area, which will be discussed later. For the present, we focus on the pragmatic approach to modeling such flows in a scale that is suitable and practical. Such a scale lies between the equipment scale, D and the dispersed phase scale d so that large scale flow structures in equipment can be captured, doing away with such simplifying assumptions as homogeneous condition inside a process vessel, which is hardly met in any real situation.

The following figure illustrates the conceptual basis of such volume averaging on a scale larger than the dispersed phase scale and specifically illustrates the progressive loss of information with increasing averaging scale. . Let the red region illustrate the dispersed phase in a continuous blue region. The yellow circle (whose size can be varied) illustrates the local volume averaging scale. At the center of yellow dot we ascribe an average value for each of the two phases, blue and red. The center of the yellow dot is continuously moved around each point inside the larger vessel, thus computing an average value at each point. Depending on the size of the yellow dot, one can observed the progressive loss of resolution of the interface. If the yellow dot fills completely the entire vessel, then we get a *single* value for the red and blue region for the entire vessel and that is the current status of assumption in most process equipment (a homogeneous, well mixed fluid). One would like to get some idea of large scale structures inside such vessels, without being concerned about the precise location of each dispersed phase particle of droplet. Starting with the early work of Whitaker and Slattery, the averaging process has been treated mathematically by several authors [16-18].

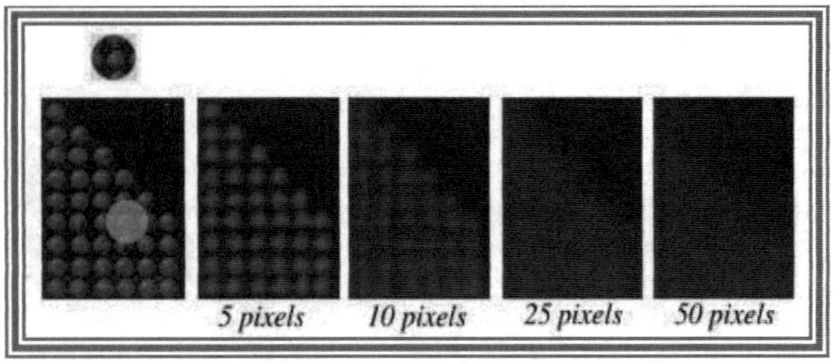

Figure 1. Volume averaging concept.

B. *Summary of volume averaged equations*

The volume averaged continuity, momentum and species conservation equations that are typically used in CFD solvers such as ANSYS-CFX, Fluent etc are summarized below.

$$\frac{\partial(\gamma_\alpha \rho_\alpha)}{\partial t} + \nabla \cdot (\gamma_\alpha \rho_\alpha \boldsymbol{U}_\alpha - \Gamma_\alpha \nabla \gamma_\alpha) = 0 \quad (1)$$

$$\frac{\partial(\gamma_\alpha \rho_\alpha \boldsymbol{U}_\alpha)}{\partial t} + \nabla \cdot \{\gamma_\alpha [\rho_\alpha \boldsymbol{U}_\alpha \otimes \boldsymbol{U}_\alpha - \mu_\alpha (\nabla \boldsymbol{U}_\alpha + \nabla \boldsymbol{U}_\alpha^T)]\} = \gamma_\alpha (B_\alpha - \nabla p) + M_\alpha \quad (2)$$

$$\frac{\partial(\gamma_\alpha \rho_\alpha Y_{i\alpha})}{\partial t} + \nabla \cdot \{\gamma_\alpha [\rho_\alpha \boldsymbol{U}_\alpha Y_{i\alpha} - \Gamma_{i\alpha} (\nabla Y_{i\alpha})]\} = \sum_{\beta=1, \beta \neq \alpha}^{N} \dot{m}_{\alpha\beta}^{i} \quad (3)$$

Here α, β represents the two phases and $(\gamma_\alpha, \boldsymbol{U}_\alpha, Y_{i\alpha})$ represent the volume fraction, velocity and concentration of species i in each phase α

C. *Need for scale invariant closure models*

The highlighted quantities need to be modeled by closure relations to capture the interphase exchange of momentum, mass and species transfer. It is essential that these closure relations capture the transport processes on the dispersed phase scale in a manner that remains scale invariant up to the size of the vessel, if reliable prediction of large scale structures and their effect on other transport processes such as heat and mass transfer are to be captured and predicted reliably. This is the most serious challenge facing the multiphase flow field. Two alternative approaches in developing such closure relations are the following: (a) experimental approach and (b) direct numerical simulation using representative number of dispersed phase entities to adequately represent meaningful averages. The problem is akin to the process of building rheological models to describe non-Newtonian fluids or turbulence models to close the problem created by time averaging. In both cases theoretical analysis has been brought to bear along with experimental observations to build models that are realistic as well as theoretically consistent. In addressing this problem in a consistent manner, multiphase flows is perhaps the least advanced. Lots of empiricism is invoked and very simple models such as drag laws on single particles are extrapolated and empirically modified to account for the effect of multi-particle effects.

D. *Packed Column Hydraulics*

Packed columns are commonly used as separation devices in distillation and gas absorption. These consist of a cylindrical vessel with some packings that have high porosity and high surface area per unit volume of packing and a gas/vapor – liquid contacting in a coutnercurrent fashion as shown in the accompanying figure.

Figure 2. Schematic of a Packed column

Traditional design approach assumes a simple one-dimensional flow, without any axial dispersion and lateral homogeneity in radial and circumferential directions yielding a simple quadrature to be evaluated in the axial direction. The simple design equation is,

$$Z = \frac{G}{K_G a P} \int_{y_1}^{y_2} \frac{dy}{y - y^*} \quad (6)$$

The work presented here has been supported on a continuing basis by the Natural Sciences and Engineering Research Council of Canada. We acknowledge the contributions of various graduate students and post doctoral fellows who worked on these projects.

In reality, there are important variations in porosity in the radial direction leading to accumulation of downward flowing liquid near the outer wall of the container, thus degrading the performance of the column. In this case the dynamics of a stationary solid phase (the packing) and two mobile fluid phases (gas and liquid) are to be studied. Their interaction and effect on the mass transfer is also to be investigated. The framework remains the same as given by the volume averaged equations (1-3). However, the closure relations are quite different, the details of which are given elsewhere [1-3].

The CFD model Eq (3) in fact reduces to the one dimensional model given by Eq (6), provided uniform one dimensional flow is assumed. However the challenge here is in the development of scale invariant closure models for the hydrodynamic drag and interface mass transfer. For example, the overall mass transfer coefficient $K_G a$ is a function of local flow conditions and it is often back calculated from Eq. (6) by measuring the performance in a known, small diameter column, which nevertheless does not have uniform, homogeneous condition on the pore scale. Hence any measurement of $K_G a$ is not truly reflective of only the interface phenomena on the dispersed phase scale, but influenced by the inhomogeneous flow conditions. Hence it is difficult to get a scale invariant closure relation. However, using one set of models we have achieved reasonable agreement for both hydraulic and mass transfer behavior as shown in the figures below.

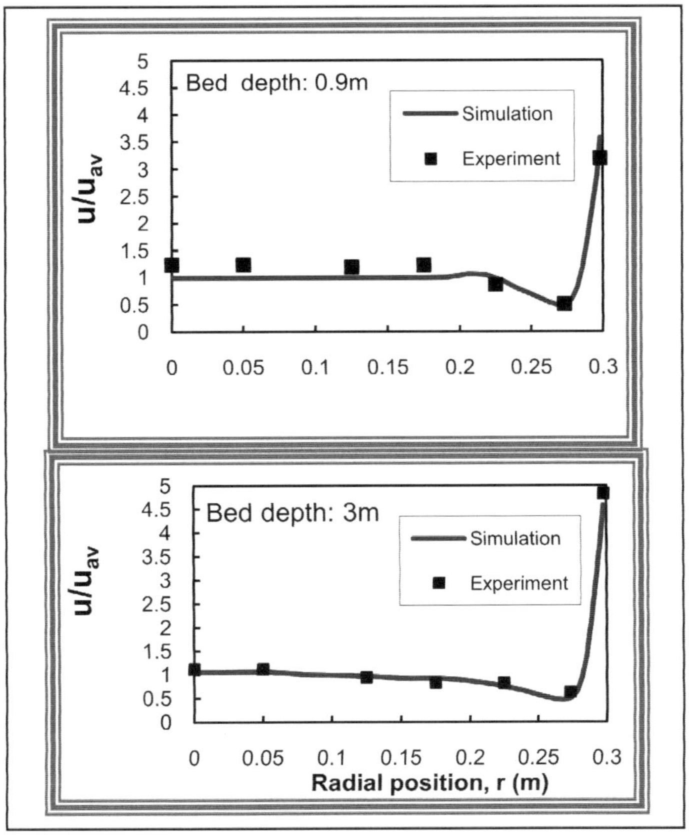

Figure 3. Wall flow in packed column by CFD simulation. Notice that the degree of wall flow increses with increasing depth of the column. uniform inlet distribution, water/air, L=4.78 kg m^{-2} s^{-1}, G=0.75 kg m^{-2} s^{-1}

Fig. 3 shows good agreement between experiments and CFD simulation for the development of wall flow.

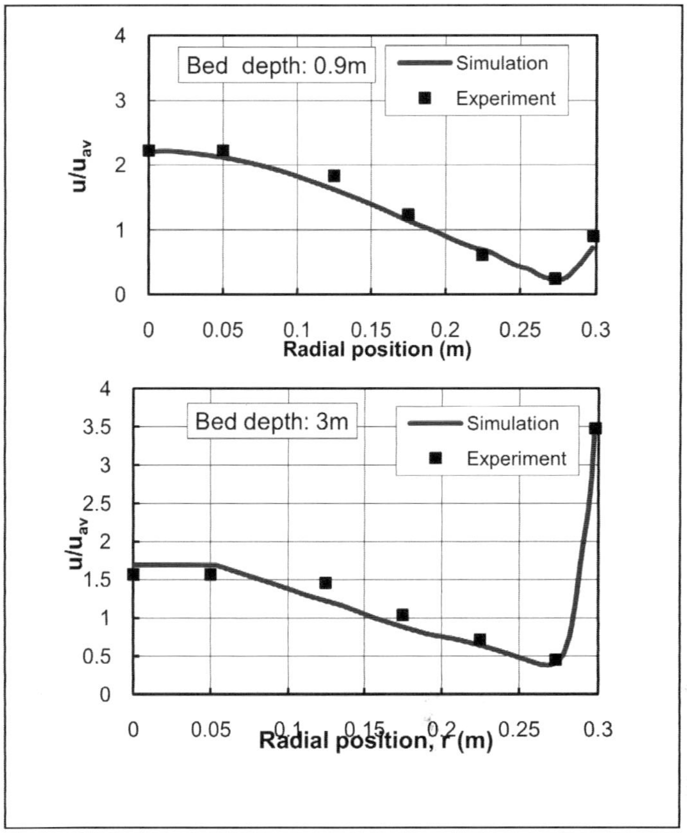

Figure 4. Wall flow in packed column by CFD simulation. Notice that the degree of wall flow increses with increasing depth of the column. 43% inlet distribution, water/air, L=4.78 kg m^{-2} s^{-1}, G=0.75 kg m^{-2} s^{-1}

Similarly Fig 4 shows good agreement between CFD simulation and experiments for the case when the outer distributors have been plugged to create a deliberate maldistribution in the packed tower. Wall flow develops much slower. The mass transfer efficiency of such columns as captured by the height equivalent of a theoretical plate (HETP) is shown in Fig 5.

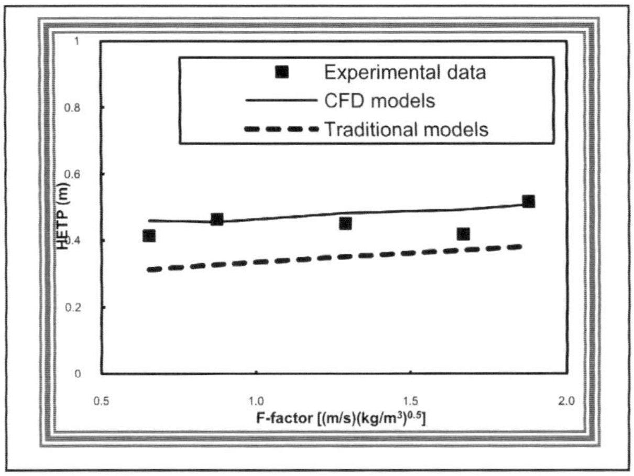

Figure 5. HETP from CFD simulation as compared to traditional algbraic empirical models

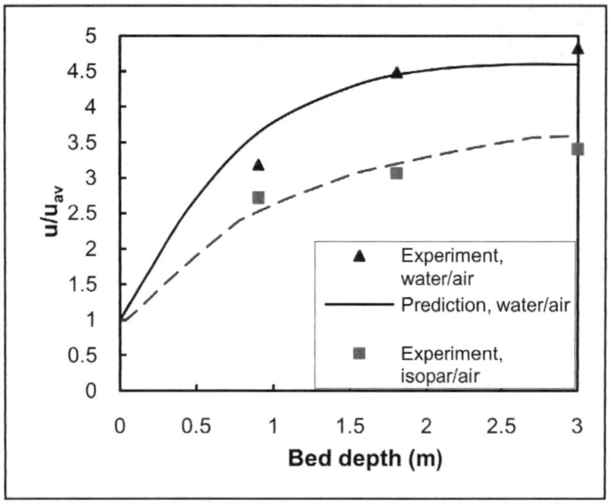

Figure 6. Developing wall flow from CFD simulation as compared to experimental measurements.

Fig. 6 shows the development of wall flow with packed bed height. By about 3 m, it reaches a fully developed or equilibrium state near the wall zone.

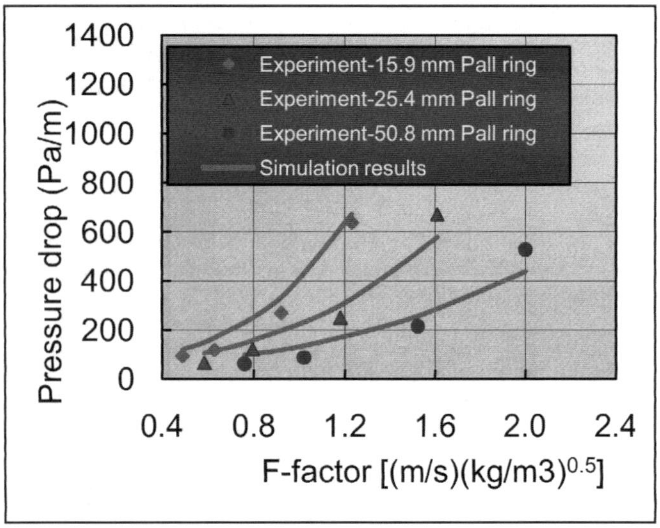

Figure 7. FRI data on pressure drop System: C_6/C_7, uniform inlet, Diameter of column: 1.22 m, 165.5 kPa.

While the results shown in Fig. 3-6 were obtained in our own laboratory using a smaller diameter column, the comparison between simulation and experiments shown in Fig. 7 are from Fractionation Research Inc. for three different sizes of Pall rings. The pressure drop models tuned with our own data are now used to predict the performance of the FRI column and the agreement is quite acceptable.

III. A Geometry based Model

In the geometry based models, the actual geometry of the packed bed is reconstructed using computer simulation. The details of the Pall rings and any other shape of the packing element is captured. Details are given in Ref. [7].

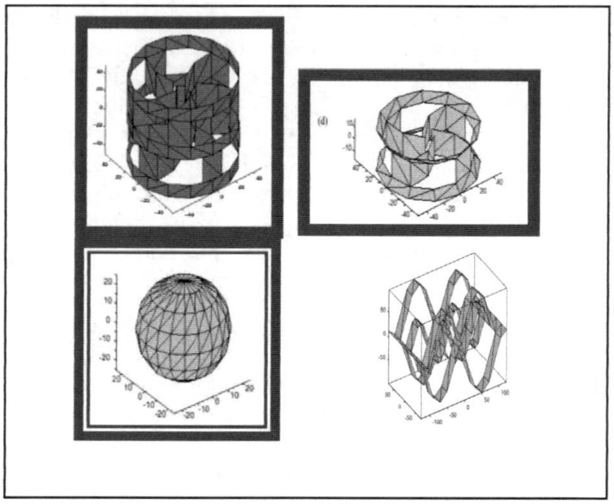

Figure 8. Geometrical details of Pall rings, mini ring, sphere, super Raschig ring.

The centroid of every particle is tracked and placed in a cylindrical vessel so that there is no overlapping. This is accomplished using collision detection algorithms. The result is a packed geometry like the one shown in Fig. 9.

Figure 9. Two thousand Pall rings packed in a cylindrical vessel.

Porosity variations can them be calculated along the radial, circumferential and axial directions by sampling a volume that is of the order of the packing element size. One such porosity variation plot is shown in Fig 10.

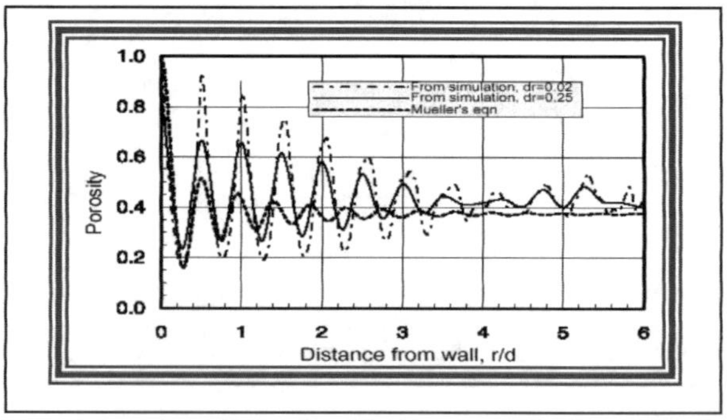

Figure 10. Porosity variation of a packed column with spheres.

Once a packed geometry is constructed in a computer, a cell based flow model is developed. Fig 11. shows the cell structure. Thin film model for flow along the inclined walls of individual packings are used to determine the flow direction and rate of liquid that is dripping along the inclined walls. Similarly the drag of air flow upwards around thin plates of arbitrary orientation are modeled to determine the pressure drop.

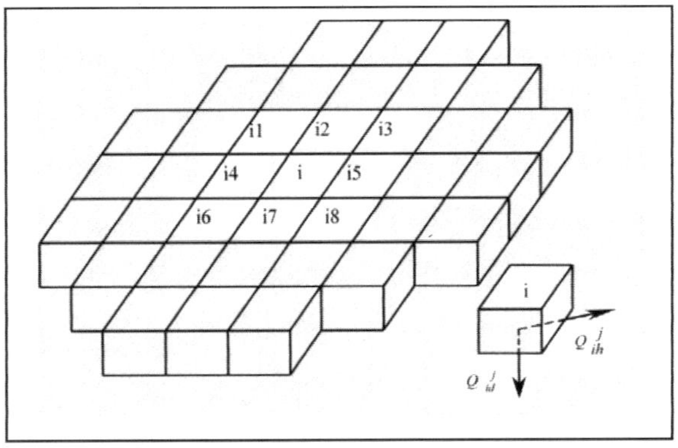

Figure 11. Cell model for hydraulics.

The liquid flow profile is then determined from a standard commercial distributor near the top (Fig 12a) and how the flow is re-distribution is determined using the cell model as shown in Fig. 12b. Details are given in [8].

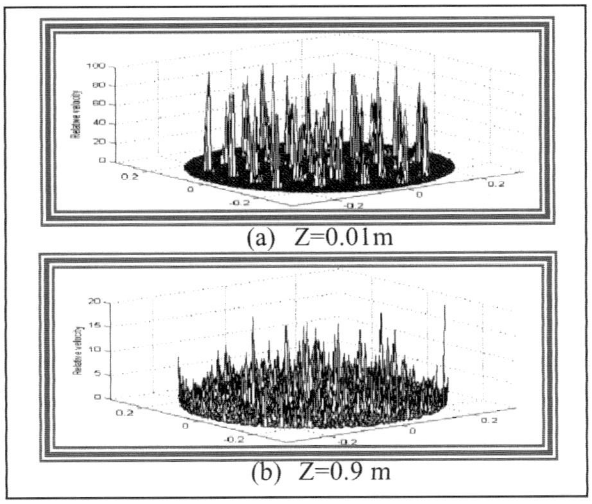

Figure 12. Cell model for hydraulics

Similar simulations can also be done for structured packings using the cell model approach. Such results will and the optimal design procedures will be presented at the conference.

IV. Conclusions

We have developed two classes of modeling approaches, one based on the interpenetrating continua model approach of CFD and the other based on detailed modeling approach coupled with cell based model. Both have as main objective, to capture the large scale flow structures that form inside packed columns and their effect on the other transport processes. Clearly closure models are needed and they are developed in such a way, that packing specific constants are minimized in the model. Although a certain degree of empiricism still exists in this approach, hopefully the range of applicability is much broader than the traditional algebraic models used in the literature.

References

[1] **F. H. Yin,** M. Song, A. Afacan, K. Nandakumar and K. T. Chuang, CFD Modeling of Mass Transfer in Randomly Packed Distillation Columns, *Ind. & Eng Chem Research* **39**:(5) (2000) pp 1369-1380.

[2] Wen, X., A. Afacan, K. Nandakumar and K .T. Chuang, A generalized approach to pressure drop predictions in packed beds based on detailed geometry of the packing, *J. Institution of Engineers Singapore,* Chem. Eng. Issue **43** (2003) pp 6-13.

[3] **F. H. Yin,** M. Song, A. Afacan, K. Nandakumar and K. T. Chuang, CFD Modeling of Mass Transfer in Randomly Packed Distillation Columns, *Ind. & Eng Chem Research* **39**:(5) (2000) pp 1369-1380.

[4] **Yin, F. H.**, A. Afacan, K. Nandakumar and K. T. Chuang, Liquid Holdup Distribution in Packed Columns: Gamma Ray Tomography and CFD Simulation, *CEP* **41** (2002) pp 473-483.

[5] **Yin, F. H.**, Z. C. Wang, A. Afacan, K. Nandakumar and K. T. Chuang, Experimental studies of liquid flow maldistribution in random packed column, *Can. J. Chem. Eng.* **78** (2000) pp 449-457.

[6] **Yin, F. H.**, A. Afacan, K. Nandakumar and K. T. Chuang, CFD Simulation and Experimental Study of Liquid Dispersion in Randomly Packed Metal Pall Ring, *Trans. Instn. Chem. Engrs.: part A. Chem. Eng. Res. & Des.* **80** (2002) pp 135–143.

[7] K. Nandakumar, Y. Shu and K .T. Chuang, Predicting geometrical properties of randomly packed beds from computer simulation, *AIChEJ.* **45** (1999) pp 2286-2297.

[8] Wen, X., Y. Shu, K. Nandakumar and K. T. Chuang, Predicting the liquid flow in randomly packed beds from computer simulations, *AIChEJ.* **47** (2001) pp1170-1179.

[9] Wen, X., A. Afacan, K. Nandakumar and K .T. Chuang, A geometry based model for predicting mass transfer in packed columns, *Ind. & Eng Chem Research* **42** (2003) pp 5373-5382.

[10] Getye Gesit, K. Nandakumar and Karl T. Chuang, CFD modeling of the flow patterns and hydraulics of commercial scale sieve trays, *AIChEJ.* **49** (2003) pp 910-924.

製程研究 PRD

台灣中油 煉製研究所

Troubleshoot of Column Containing Structured Packing

Hsiu-Hsiung Hsiao

Refining & Manufacturing Research Institute
CPC Corp.,TAIWAN

March 2010

AIChE 2010 Spring National Meeting in San Antonio

Aromatic Complex Flow Diagram

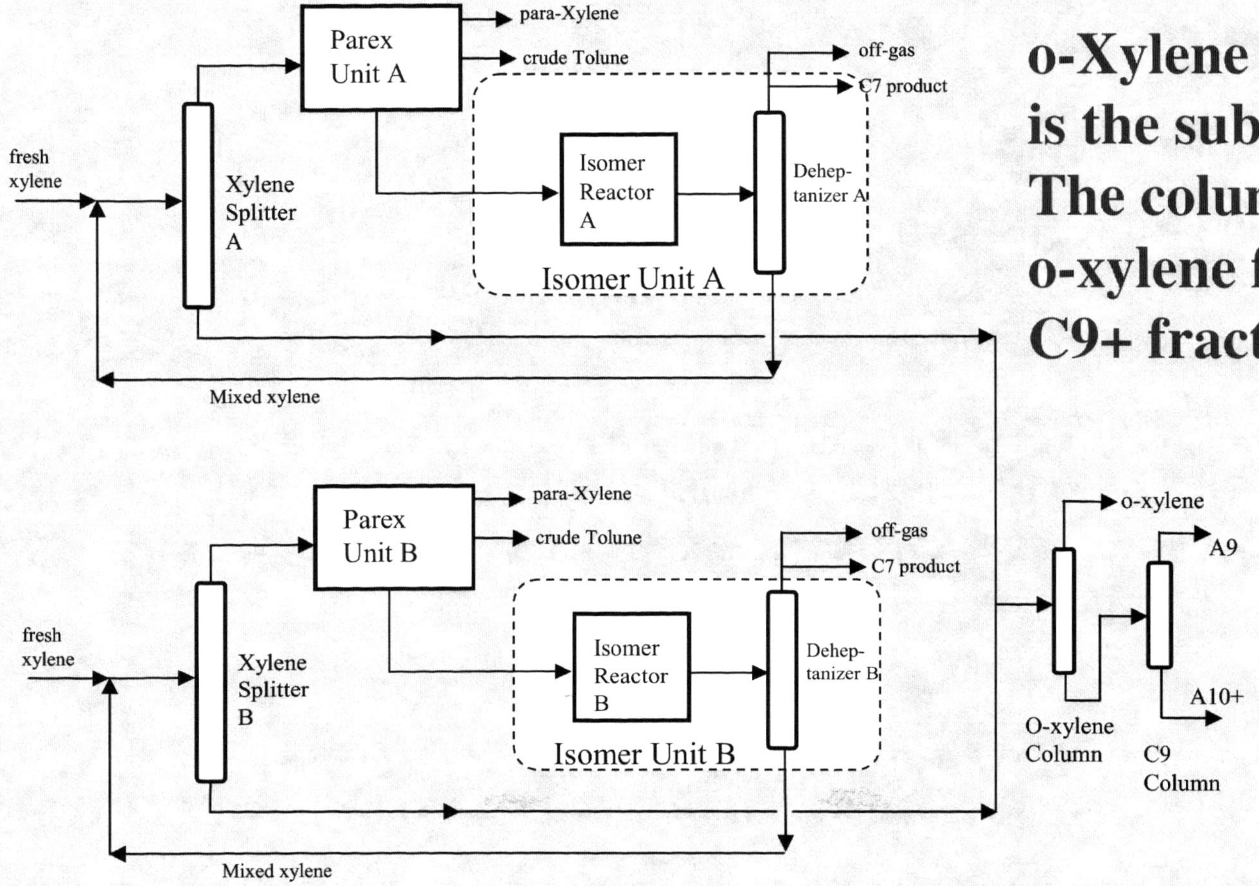

o-Xylene column is the subject column. The column separate o-xylene from heavier C9+ fraction.

Reason of revamp

- Capacity drop due to the more stringent specification on impurity ispropyl benzene content in o-xylene product, the B.P. difference between them is only 14°F .
- Old specification : o-xylene 98.5wt% in top and o-xylene 1.5% in bottom.
- New specification : o-xylene 98.5wt% & ispropyl benzene 0.3 wt% maximum in top and o-xylene 1.5wt% in bottom.

Revamp from tray to structured packing

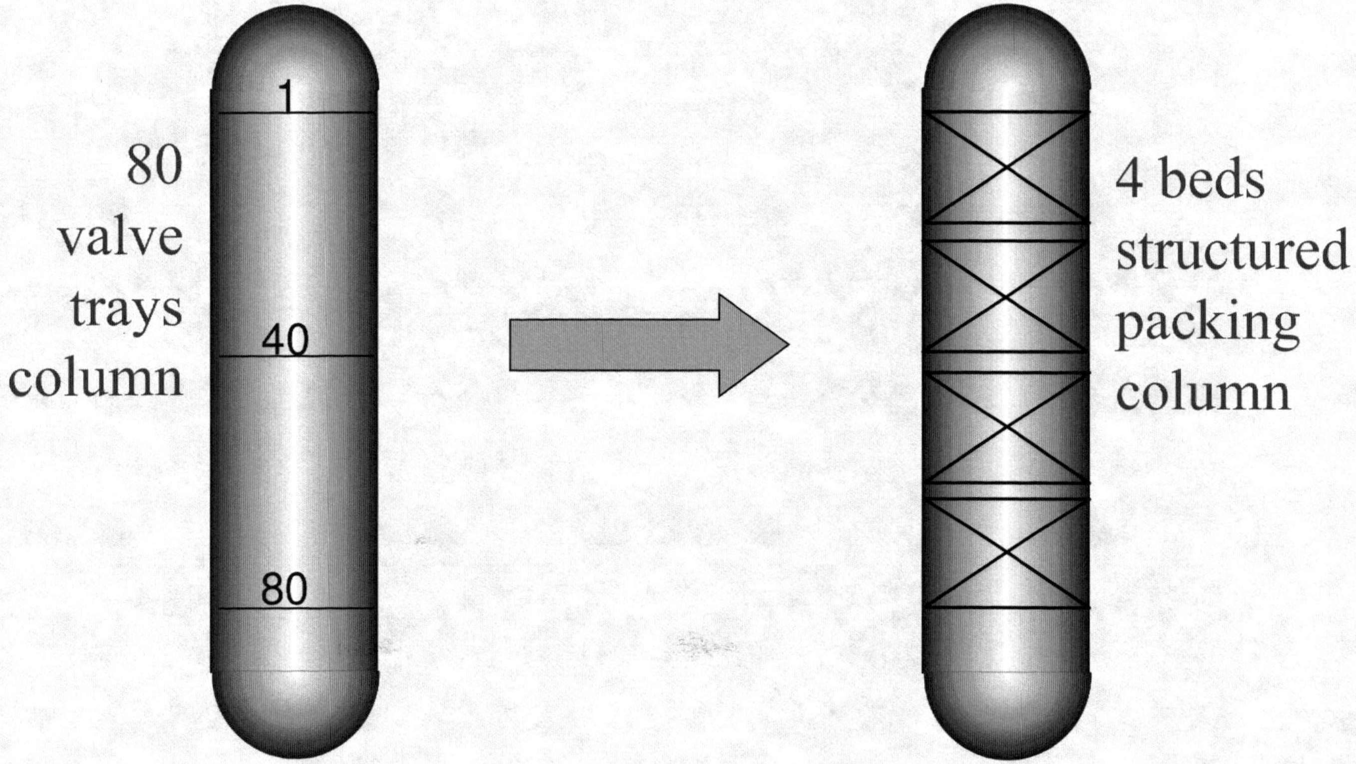

Performance after 1st install of packing

- Performance was bad , top and botom could not meet specification simultaneously, It seems the symptom of not enough theoretical tray number generated by packing.
- Change liquid collector from vane type to chimney tray type, but performance was not improved. Liquid entrainment was not the cause of bad performance.

Test data at two column loading

Column section bounded by two temp. measure point (T11 means temp. measure point at location corresponds to original tray 11)	Temp. Difference between two temp. measure points at lower column loading (°F)	Temp. Difference between two temp. measure points at higher column loading (°F)
T11 ~ T1	2.6	2.2
T30 ~ T11	4.4	8.6
T37 ~ T30	11.8	10.3
T43 ~ T37	7.0	8.0
T57 ~ T43	18.4	17.3
T64 ~ T57	6.3	4.2
Feed rate (GPM)	70	100
Reflux rate (GPM)	119	170
isopropyl benzene content in top (wt%)	0.3	0.29
o-xylene content in bottom(wt%)	1.7	6.2
Theoretical plate number	45	37
Capacity factor (%)	65	78

- **The isopropyl benzene content in top was almost the same in two cases.**
- **Temp. difference reflects the composition difference, the pressure effect could be neglected.**
- **Column loads in both cases were lower than flooding condition.**

Analysis of test data (1)

- Overall performance became worse as column loading increase. Simulation comfirmed it. Less theoretical plate number for higher loading case.
- The temp. difference in section T64~T57 for lower loading case was 2.2°F higher than that for higher loading case. It means the separation in stripping section became worse as column loading increase.

Analysis of test data (2)

- The temp. difference in section T30~T11 for higher loading case was 4.2°F higher than that for lower loading case. Some people will say that the separation in rectification section for higher loading case is better. But it is wrong. The correct explaination is that much more heavier components reach location T30 for higher loading case than for lower loading case. Therefore, the separation in rectification section became worse as column loading increase too.
- The source of the performance deteriorations as loading increase was believed to be liquid and vapor maldistributions caused by overly-deep packing beds.

Comparison of bed length in 1st and 2nd installation

	Bed length in 1st installation (mm)		Bed length in 2nd installation (mm)	
Rectification	bed 1	8938	bed 1	5668
			bed1A	5668
	bed 2	8938	bed2	5668
Stripping	bed3	8938	bed3	6104
			bed3A	6322
	bed4	8938	bed4	6322

Comparison of performance before and after 2nd installation

Column section bounded by two temp. measure point (T11 means temp. measure point at location corresponds to original tray 11)	Temp. Difference between two temp. measure points at lower column loading before 2nd installation(°F)	Temp. Difference between two temp. measure points at higher column loading before 2nd installation(°F)	Temp. Difference between two temp. measure points after 2nd installtion (°F)
T11 ~ T1	2.6	2.2	1.4
T30 ~ T11	4.4	8.6	6.1
T37 ~ T30	11.8	10.3	10
T43 ~ T37	7.0	8.0	2.5
T57 ~ T43	18.4	17.3	15.6
T64 ~ T57	6.3	4.2	15.6
Feed rate (GPM)	70	100	100
Reflux rate (GPM)	119	170	145
isopropyl benzene content in top (wt%)	0.3	0.29	0.3
o-xylene content in bottom(wt%)	1.7	6.2	1.4
Theoretical plate number	45	37	70
Capacity factor (%)	65	78	75

The column is now achieving original design targets.

(133e) Dr. K. Chuang - From Heavy Water to the Superfrac(r) Tray

Darius R. Remesat, Koch-Glitsch Canada LP, Calgary, AB, Canada

Presentation goes over some of Dr. K. Chuang's contributions to distillation:

1. Early developments in heavy water distillation 2. Contributions to Superfrac(r) tray development 3. Examples of successfull superfrac(r) tray installations

See more of this Session: Advances in Distillation & Absorption V
See more of this Group/Topical: Topical 8: Distillation Topical

(133f) Professor Karl Chuang Contributions to Separation Science

Mike Resetarits, Fractionation Research, Inc., Stillwater, OK

Ahmad Shariat, Fractionation Research, Inc., Stillwater, OK

Tony J. Cai, Fractionation Research, Inc., Stillwater, OK

A.Y. Ogundeji, Fractionation Research, Inc., Stillwater, OK

Few chemical engineers have spent as much time at as many frontiers as Professor Karl T. Chuang. Of his 200 refereed papers, 300 conference papers and 40 patents, approximately half regarded catalysis and the other half regarded separations. His distillation/absorption research addressed hydraulics, efficiency, modelling, CFD, intelligent design and catalytic distillation. Obviously, trays and packings were studied. Less obviously, trayed packed columns and packed trayed columns were studied. Appreciable, and important, data from the excellent University of Alberta laboratory were released to the public. This paper presents the separations-related bibliography and highlights of Professor Chuang's amazing career.

See more of this Session: Advances in Distillation & Absorption V

See more of this Group/Topical: Topical 8: Distillation Topical

AUTHOR INDEX